THE ILLUSTRATED ELECTRONICS DICTIONARY

HOWARD M. BERLIN

Delaware Technical and Community College
Stanton Campus

Charles E. Merrill Publishing Company
A Bell & Howell Company
Columbus   Toronto   London   Sydney

"No dictionary of a living tongue ever can be perfect, since while it is hastening to publication, some words are budding and some are falling away."

Samuel Johnson, 1755

This dictionary is dedicated to my children, Jessica, Mark, and Daniel, in the hope that they may develop the same respect for the preciseness of the English language that my father gave to me.

Published by Charles E. Merrill Publishing Co.
A Bell & Howell Company
Columbus, Ohio 43216

This book was set in Rockwell and Univers
Copy Editor: Robert L. Marcum
Cover Design Coordination: Cathy Watterson
Cover Photo: Joan Kramer & Associates/Photographer—Glenn Steiner
Production Coordination: Molly Kyle

Library of Congress Catalog Card Number: 85–62113
International Standard Book Number: 0–675–20451–8
Printed in the United States of America
    2   3   4   5   6   7   8   9   10—91   90   89   88   87   86

PREFACE

This illustrated dictionary is a valuable reference for practicing engineers, technicians, technical writers, secretaries, and students. The electronics explosion over the past decade has generated many new terms and brought a revival or revision of existing terms in electricity and electronics. This volume lists more than 2000 commonly used words, phrases, abbreviations, acronyms, and symbols in electricity and electronics, accompanied by more than 200 illustrations.

Terms are alphabetized letter by letter, irrespective of spaces, hyphens, and abbreviations. Greek letters are listed by their English spellings. Although a definition appears only once, synonyms and cross-references have been placed at the end of appropriate entries, and cross-references are italicized.

Following the dictionary section, you will find helpful supplementary material: rules of style in electricity and electronics, multiplier prefixes, standard quantity symbols, Greek letter symbols, and schematic symbols. Because Charles E. Merrill Publishing Co. has another volume devoted to computer definitions, *The Illustrated Computer Dictionary* by Donald D. Spencer, terms primarily associated with computer technology are minimized here.

Howard M. Berlin
Wilmington, Delaware

CONTENTS

Dictionary **1**

Rules of Style in Electricity and Electronics **177**

Multiplier Prefixes **180**

Standard Quantity Symbols **181**

Standard Unit Symbols **183**

Greek Letter Symbols **184**

Schematic Symbols **185**

a Prefix symbol for *atto* (10^{-18}).

A (1) Unit symbol for *ampere*. (2) Symbol for *gain*. (3) Abbreviation for *anode*. (4) Symbol for the area of a plane surface.

absolute temperature scale See *Kelvin scale*.

absolute value The numerical value of a number without regard to polarity or sign.

absolute zero The temperature at which all molecular motion ceases; equal to $-273.16°C$ or $-459.69°F$.

absorption loss (1) Energy loss in a transmission line that is converted to heat when radiated energy is transmitted or reflected by a material. (2) The power loss in a transmission circuit due to adjacent circuit coupling.

absorption wavemeter A type of wavemeter which consists of a calibrated tuned circuit and a resonance indicator. When it is lightly coupled to a signal source and tuned to resonance, maximum energy is absorbed from the source. See *grid-dip meter*.

AC (or ac) Abbreviation for *alternating current*.

access time The time required to communicate with a storage device. The access time is characteristic of that device.

AC coupling The coupling of one circuit to another, generally using a capacitor, whereby only the time-varying component of an electrical signal is passed through. The signal's DC, or static component, is blocked.

AC current Abbreviation for *alternating current* current; mathematically symbolized by i. Although the term is redundant and awkward, it has been frequently used to distinguish the current from AC voltage.

AC/DC An abbreviation meaning that a particular device will operate from either an AC or a DC power source.

AC generator A rotating machine that converts mechanical power into electrical power. Also called an *alternator* or *dynamo*.

AC receiver A radio receiver that operates only from an AC power line.

AC relay A relay whose coil responds to an AC voltage.

AC resistance Symbolized by *r*, the resistance between two points in an AC circuit.

active arm (leg) The electrical element of a transducer whose electrical characteristic changes in proportion to an applied stimulus.

active component An electronic component that requires one or more sources of power, such as a transistor that produces gain or switching action in a circuit. Also called an *active device* or *active element*.

active device See *active component*.

active element See *active component*.

active filter A filter network using an active device to provide gain in the filter's passband.

active transducer A transducer that has a power source whose output is controlled by an electric signal.

actual power The average value of instantaneous power taken over one cycle.

AC voltage Abbreviation for *alternating current* voltage; mathematically symbolized by *v*. Although awkward, the term has been frequently used to differentiate the voltage from AC current.

adapter A fitting that changes the terminal connections of a socket, plug, jack, or other receptacle so that other electrical connections can be made besides those of the original connector.

A/D converter (or ADC) Abbreviation for *analog-to-digital converter*.

A-display A radar display in which targets appear as vertical deflections from a horizontal time base line.

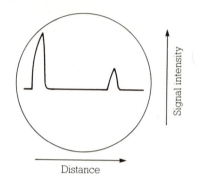

A-display

adjacent channel The frequency band immediately above or below a given band of frequencies.

adjacent channel interference An unwanted signal received on one communications channel from a source transmitting on a frequency immediately above or below the received channel.

admittance Symbolized by the letter *Y*, the ratio of the current to voltage in an electric network or branch.

aerial British term for *antenna*.

AF (or af) Abbreviation for *audio frequency*.

AFC (or afc) Abbreviation for *automatic frequency control*.

AFSK Abbreviation for *audio frequency shift keying*.

AGC (or agc) Abbreviation for *automatic gain control*.

age To store an electrical component in a specified environment until its characteristics stabilize.

air capacitor   A capacitor, usually variable, with air as the dielectric medium between its parallel plates.

alias frequency   An unwanted low frequency obtained when a periodic signal is sampled at a rate of less than one-half its frequency.

alignment tool   A small screwdriverlike tool made of nonmagnetic materials used for the adjustment of tuned circuits.

alkaline cell   Also called an *alkaline-manganese cell,* a battery cell that uses potassium hydroxide electrolyte to produce 1.5 volts. It has higher current ratings than a carbon-zinc dry cell.

alligator clip   A long, narrow, springlike clip with meshing jaws that is used with test leads to make quick, temporary connections.

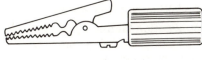

Alligator clip

all-pass network   A filter network providing phase shift proportional to frequency but not attenuating the signal over the same range.

alpha (α)   First letter of the Greek alphabet, used as the symbol for the *forward current gain from emitter to collector* of a transistor wired in a grounded-base configuration. It is the ratio of the change in collector current to the corresponding change in emitter current and is always less than 1.

alpha cutoff frequency   Symbolized by f_α, the high frequency at which the amplification factor (α) of a transistor decreases by a factor of 0.707, or 3 dB from its low-frequency value.

alpha particle   A positively charged atomic particle that has two protons and a single neutron; identical with the nucleus of a helium atom.

alternating current   Abbreviated *AC,* an electric current, such as a sinewave, that is continually varying in value as well as reversing in direction of flow at a periodic interval.

alternator   A rotating machine that uses electromagnetic induction to generate an AC voltage when its armature or field is rotated by motor, engine, hand, or other means. Its output frequency is directly proportional to the speed at which it is driven. Also called a *synchronous generator, dynamo,* or *ac generator.*

AM (or am)   Abbreviation for *amplitude modulation.*

ambient   The immediate surrounding environment.

American Wire Gage   Abbreviated *AWG,* a system of specifying the wire sizes (diameters) used in the United States. The larger the gage number, the smaller the diameter.

ammeter   An instrument that measures either AC or DC current flow.

amp   (1) Abbreviation for *ampere.* (2) Slang for *amplifier.*

ampacity   A power cable rating of its current-carrying capacity in amperes.

amperage   The amount of current in amperes.

ampere Abbreviated *A*, or *amp*, the unit of electrical current equal to the flow of electric charge at the rate of 1 coulomb per second; equal to the current flowing through a 1-ohm resistor with a voltage drop of 1 volt across it.

Ampere, Andre Marie A French physicist and mathematician (1775–1836) who amplified the work of Oersted on the relationship of electricity and magnetism, and after whom the unit of current was named.

ampere-hour Abbreviated *Ah*, a measure of electric charge applied in particular to rating storage batteries. It is defined as the charge transferred by a current of 1 ampere flowing for 1 hour, and is equal to 3 600 coulombs.

Ampere's law The line integral of the magnetic field vector around a single closed path equals the current enclosed.

Ampere's rule Following from Ampere's law, if a current flows in a wire directly towards an observer, then the magnetic field or flux circulates counterclockwise around the wire. Also known as the *right-hand screw rule:* If the thumb of the right hand points in the direction of the current flow, the fingers of the right hand encircle the wire in the direction of the magnetic field or flux.

ampere-turn Abbreviated *At,* the unit of magnetomotive force that is the result of the force developed by passing a current of 1 ampere through a 1-turn coil of wire. Equal to 1.26 gilberts.

amplidyne Acronym for amplifier dyne, a rotating magnetic amplifier that consists of a DC motor and generator with special windings and brush connections to give power

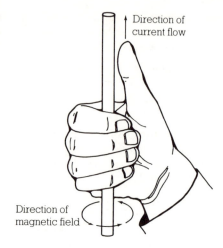

Direction of current flow

Direction of magnetic field

Ampere's rule (right-hand screw rule)

amplification from small changes in power input to the field coils.

amplification The process of increasing the intensity (current, power, or voltage) of a signal.

amplification factor Symbolized by the Greek letter μ, a dimensionless parameter of a vacuum tube defined as the ratio of a small change in plate (anode) voltage to the corresponding change in grid voltage while the plate current is held constant.

amplifier A circuit using an active device, such as a tube, transistor, or integrated circuit, to increase the intensity of a signal without changing the shape of the waveform.

amplitude (1) The value of a varying quantity, such as voltage, at a specified instant of time. (2) Of a complex quantity, the magnitude of its real and imaginary components.

amplitude distortion See *frequency distortion.*

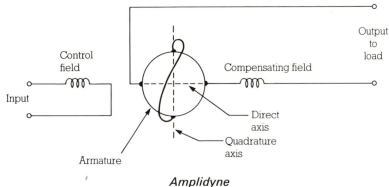

Amplidyne

amplitude gate A circuit that transmits only the portion of an input signal lying between two amplitude values.

amplitude modulation Abbreviated *AM*, a type of modulation in which the amplitude of a carrier signal is varied about its normal value in accordance with the shape of the intelligence (modulating signal) to be transmitted. The resultant waveform has a frequency component that is equal to the carrier, plus components equal to the sum and difference frequencies of the carrier and modulating signals.

analog Also spelled analogue, the representation of numerical quantities by physical variable, such as AC voltage or temperature. The opposite of digital.

analog data Data represented in continuous form rather than as discrete values.

analog panel meter A conventional meter whose value is indicated by a pointer needle moving over a calibrated scale.

analog signal (1) A signal whose magnitude represents information content. (2) A nondigital signal.

analog switch A solid-state device that either transfers or blocks an analog signal without distortion by applying a control signal.

analog-to-digital converter Abbreviated *A/D converter* or *ADC*, a device in which the magnitude of an analog signal is converted to its equivalent binary value.

Anderson bridge A six-arm modification of a Maxwell-Wein bridge, used to measure self-inductance in terms of capacitance and resistance, with the bridge balance being independent of frequency. *(See figure on page 6.)*

AND gate A digital device whose output is at logic 1 when all of its inputs are at logic 1; otherwise its output is at logic 0.

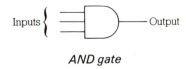

AND gate

anechoic Nonreflective; producing no echos.

angle of radiation See *radiation angle.*

5

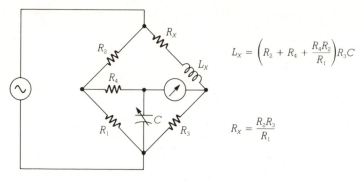

$$L_X = \left(R_2 + R_4 + \frac{R_4 R_2}{R_1} \right) R_3 C$$

$$R_X = \frac{R_2 R_3}{R_1}$$

Anderson bridge

angular frequency   Symbolized by the Greek letter ω, frequency expressed in radians per second rather than in hertz; equal to 2π times the frequency in hertz.

anion   A negative ion.

ANL   Abbreviation for *automatic noise limiter*.

anode   (1) Symbolized by the letters A or P, the positive terminal of an electron tube. (2) British term for the plate of an electron tube. (3) In a semiconductor diode, the terminal towards which forward current flows. (4) The positive terminal of a cell or battery.

antenna   A device that either radiates or absorbs radio or electromagnetic waves. Also called an *aerial.*

antilog   See *antilogarithm.*

antilogarithm   (1) The number from which a given logarithm is derived. Also called an *antilog.* (2) The inverse operation of a logarithm.

antiresonant circuit   See *parallel-resonant circuit.*

aperiodic   Not periodic.

arc   (1) A discharge of electricity through a gas. (2) A segment of any curve. (3) In mathematics, the inverse; especially applied to trigonometric and hyperbolic functions.

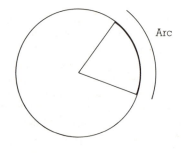

Arc

arm   A branch of a circuit, particularly that of a bridge network.

armature   (1) In an electric generator, the coil in which the electromotive force is induced. In DC generators the armature is a rotor, but in alternators it may be either a stator or rotor. (2) The moving element of an electromechanical device, such as a relay.

Armstrong, Edwin Howard   An American engineer (1890–1954) who made many contributions to the development of radio, including the inventions of the superheterodyne receiver and frequency modulation.

Armstrong oscillator A type of parallel-resonant, tuned-feedback oscillator built around either a vacuum tube or a transistor. In a transistor, feedback is achieved through its base-collector capacitance while the collector load is tuned slightly higher above the output frequency so that the load appears inductive.

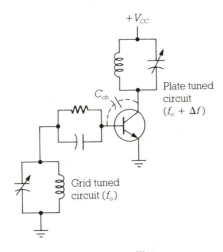

Armstrong oscillator

arrester A protective device used to provide a bypass path directly to ground for lightning discharges that strike an antenna or other conductor. Also called a lightning arrester.

ASCII Acronym for American Standard Code for Information Interchange.

aspect ratio The frame width-to-height ratio of a television screen. In the United States, it is 4:3.

astable (1) Not stable. (2) Having no stable output states.

astable multivibrator A multivibrator circuit whose output voltage alternates between two unstable states at a constant frequency and duty cycle without the used of exter-

nal triggering. Also called a free-running multivibrator or clock.

aster rectifier A circuit that uses 12 or more rectifiers with a conducting period of 30 electrical degrees plus the commutating angle.

asymmetrical sideband See *vestigial sideband*.

asynchronous Not synchronous.

asynchronous machine An AC machine whose speed is not proportional to the power line frequency.

attenuation A reduction in signal intensity, usually expressed in decibels.

attenuator A device or circuit used to reduce the intensity of a signal without distortion.

atto Abbreviated *a*, prefix for 10^{-18}.

audio (1) Pertaining to phenomena that involve frequencies in the range of human hearing. (2) Slang term for *sound*.

audio amplifier An amplifier capable of amplifying audio frequency signals. Also called an audio frequency amplifier.

audio frequency Abbreviated *AF*, a frequency that is audibly detected by the human ear, generally in the range from 16 Hz to 20 kHz. Also called *sound* or *sonic frequency*.

audio frequency choke An inductor used to block the passage of audio frequency currents.

audio frequency shift keying Abbreviated *AFSK*, a form of radioteletype transmission in which the RF carrier is transmitted continuously while pulses are transmitted by frequency-shifted audio tones.

audion The original three-element vacuum tube developed by Lee de Forest.

audio signal An electric signal that has a frequency within the audio frequency range.

automatic frequency control (1) Abbreviated *AFC*, a circuit that keeps the output frequency of an oscillator constant. (2) A circuit that keeps a superheterodyne receiver accurately tuned to a given frequency by controlling its local oscillator. (3) In a television receiver, a circuit that locks the frequency of the sweep oscillator onto the synchronization pulses of the received signal.

automatic gain control Abbreviated *AGC*, a circuit that automatically changes the gain of a device so that the output signal level remains essentially constant with changes in the input signal intensity.

automatic noise limiter Abbreviated *ANL*, a circuit that clips off all noise peaks that are stronger than the largest desired peak in the signal being received.

automatic volume control Abbreviated *AVC*, an automatic gain control that maintains the output volume of a radio receiver essentially constant despite changes in input signal strength.

autopolarity Automatic interchanging of connections to a digital meter when the polarity is incorrect.

autoranging Automatic switching of a multirange instrument from its lowest range to the next highest range without the meter reading exceeding the maximum scale limit.

autotransformer A power transformer that has one continuous tapped winding. Part of the tapped winding serves as the primary while all of it serves as the secondary, or vice versa.

avalanche breakdown The non-destructive breakdown of a reverse-biased p-n junction when the electric field becomes strong enough to ionize a few atoms of the semiconductor material.

avalanche diode A silicon diode having a high reverse-to-forward resistance ratio until avalanche breakdown occurs. After breakdown, the voltage drop across the diode is essentially constant. Also called a *breakdown diode.*

AVC (or avc) Abbreviation for *automatic volume control.*

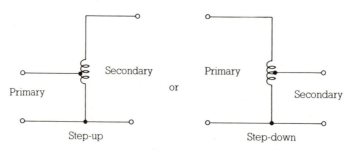

Autotransformer

average value (1) The value obtained by taking the sum of N values and dividing this sum by the number of values. (2) For an AC signal, the value of the area underneath the curve divided by the signal's period.

AWG Abbreviation for *American Wire Gage.*

Ayrton-Perry winding The winding of two wires in parallel but opposite directions to give better cancellation of magnetic fields.

Ayrton shunt A resistive voltage divider string, connected in parallel with a galvanometer in order to provide variable meter sensitivity as well as meter damping.

azimuth The distance in degrees (or mils) in a clockwise direction from the north point.

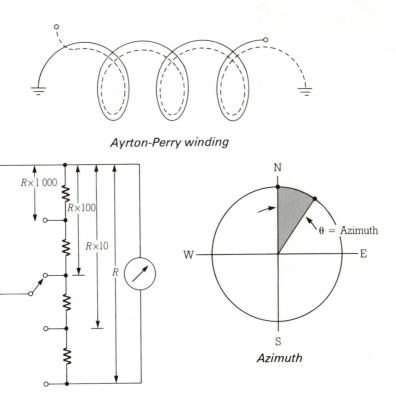

Ayrton-Perry winding

Ayrton shunt

Azimuth

B (1) Mathematical symbol for *magnetic flux density, susceptance*.
(2) Abbreviation for *base, bel*.
(3) Letter designation for *battery* on schematic diagrams.

babble In a communications channel, interference in the form of crosstalk from many other channels.

back bias See *reverse bias*.

back current Reverse current.

background noise Noise heard along with desired signals.

baffle A reflective shield or partition used to guide some form of radiation or flux along a desired path.

balanced A network, transmission line, or signal in which all components are symmetrical with respect to a common reference point, such as ground.

balanced amplifier An amplifier, such as a push-pull amplifier, in which there are two identical signal branches connected to operate in phase opposition, with input and output connections each balanced to ground.

balanced bridge A bridge circuit adjusted so that the output voltage is zero.

balanced circuit A circuit whose two sides are electrically the same and symmetrical with respect to a common reference point, such as ground.

balanced converter A balun.

balanced line A transmission line having two conductors with voltages at any point equal in magnitude but opposite in polarity with respect to ground.

balanced modulator A push-pull amplitude modulator that completely suppresses the carrier signal but produces an output that consists of only the upper and lower sidebands. Also called a *balanced ring modulator*.

balanced network A network that has equal input and output impedances.

balanced ring modulator A modulator that uses diodes to completely suppress the carrier signal but produces an output that consists of only the upper and lower sidebands. Also called a *balanced modulator*.

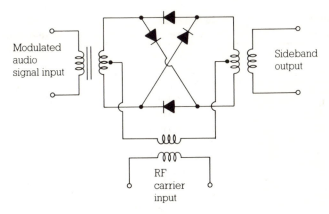

Modulated audio signal input

Sideband output

RF carrier input

Balanced ring modulator

balanced termination A two-terminal load in which both terminals present the same impedance to ground.

balanced transmission line See *twin lead.*

balun A device or transformer used to match a balanced transmission line to an unbalanced line in addition to matching impedances.

banana jack A jack that accepts a banana plug.

banana plug A plug that has a spring-metal tip which resembles a banana, used on test leads.

Banana plug

band A range of frequencies between two given limits. The range of useable frequencies is divided into 10 bands from extremely low frequency (ELF, below 300 Hz) to extremely high frequency (EHF, 30–300 GHz). The VHF and UHF bands are further divided into single letter bands. *(See table on page 12.)*

band-elimination filter. See *notch filter.*

bandpass See *passband.*

bandpass amplifier An amplifier that passes a definite band of frequencies with an essentially uniform response.

bandpass filter A filter network, either active or passive, that allows signals between two frequencies to pass, but rejects signal frequencies above and below this range.

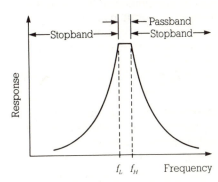

Bandpass filter response

Band Name	Frequency Range	Wavelength
VLF	10–30 kHz	10–100 km
LF	30–300 kHz	1–10 km
MF	300 kHz–3 MHz	100–1 000 m
HF	3–30 MHz	10–100 m
VHF	30–300 MHz	1–10 m
UHF	300 MHz–3 GHz	10–100 cm
SHF	3–30 GHz	1–10 cm
EHF	30–300 GHz	1–10 mm
P	225–390 MHz	76.92–133.3 cm
L	390 MHz–1.55 GHz	19.35–76.9 cm
S	1.55–5.2 GHz	5.77–19.35 cm
C	3.9–6.2 GHz	4.84–7.69 cm
X	5.2–10.9 GHz	2.75–5.77 cm
K	10.9–36 GHz	8.33–27.5 mm
Q	36–46 GHz	6.52–8.33 mm
V	46–56 GHz	5.36–6.52 mm

band-reject filter   See *notch filter.*

band-stop filter   See *notch filter*

bandwidth   Abbreviated *BW*, the frequency range, generally taken between two frequencies, where the response is 3 dB less than the average passband response.

bank   A number of similar devices, such as resistors, connected together for use as a single device.

barrel distortion   In a camera or image tube, the distortion which results in a monotonic decrease in radial magnification in the reproduced image away from the axis of symmetry of the electron optical system. In TV receivers, barrel distortion makes all four sides of the raster curve out like a barrel.

barrier voltage   The minimum voltage required for conduction through a p–n semiconductor junction.

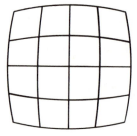

Barrel distortion

base   Abbreviated *B*, the control terminal of a bipolar transistor; two of three terminals in a unijunction transistor.

baseband   The band of frequencies occupied by all the transmitted signals that are used to modulate a particular carrier.

base line (or baseline)   The line on an oscillographic display that the sweep produces in the absence of an input signal.

bass Sounds in the low-frequency audio range, generally those below that of middle C (256 Hz).

battery A DC voltage source consisting of one or more cells which converts chemical, solar, or thermal energy into electrical energy.

baud Abbreviated *Bd*, a unit of signaling speed in telecommunications, equal to one element per second.

bazooka A balun.

bead A ceramic, glass, or plastic insulator with a hole through its center, used to support the inner conductor of a coaxial line in the exact center of the line. Ferrite beads are placed on high-frequency wire connections of circuits to suppress parasitic oscillation by providing inductance.

bead thermistor A thermistor that consists of a small bead of semiconductor material placed between two wire leads, used mainly in the measurement of temperature and microwave power levels.

beam finder A switch on some cathode-ray oscilloscopes which places the beam trace on the screen regardless of the vertical and horizontal amplifier and intensity controls.

Bd Abbreviation for *baud*.

beat frequency Either the sum or difference frequency output of a mixer stage when two signals of different frequencies are heterodyned.

beat frequency oscillator Abbreviated *BFO*, an oscillator whose output is heterodyned with another signal to produce a beat frequency.

bel Abbreviated *B*, the fundamental unit of sound level or gain, equal to the common (base 10) logarithm of the ratio of two power levels, so that 1 bel is equal to a power gain of 10.

bell wire Cotton or plastic covered No. 18 copper wire, used primarily for doorbell and thermostat connections inside a house.

Bessel filter A type of active low-pass or high-pass filter whose rolloff is less than a Butterworth filter, but whose phase shift is linear with frequency. Because the filter is highly damped, transient responses are passed with a minimum of distortion.

beta Symbolized by the Greek letter β, the short-circuit, common-emitter current gain of a transistor, equal to the ratio of the collector current to the base current. Also called *current transfer ratio*.

BeV Abbreviation for *billion electronvolt*.

BFO (or bfo) Abbreviation for *beat frequency oscillator*.

B-H curve A characteristic curve showing the relationship between the magnetic induction (B) and the magnetizing force (H) of a magnetic material. As a consequence, it is also the manner in which the permeability of a material varies with flux density. Also called *magnetization curve*.

bias (1) A steady-state force, magnetic field, or DC voltage applied to a system in order to establish a reference level or to determine the range of operation. (2) To apply a bias to.

bifilar transformer A transformer in which the wires for the primary and secondary windings are wound side by side to give a high degree of coupling.

13

bifilar winding A winding method such that two insulated wires are wound side by side with the currents through them traveling in opposite directions in order to obtain the maximum coupling between the two circuits.

binding post A manually-turned bolt and nut terminal mounted on a chassis or panel, used for making temporary electrical connections.

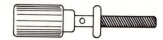

Binding post

Biot-Savart's law A law stating that the magnetic field intensity produced by a current flowing in a straight wire at any radical distance from the wire is proportional to the current flowing in the wire and inversely proportional to the radial distance.

bipolar (1) Having two poles, polarities, or directions. (2) Having majority and minority carriers present.

bipolar device A semiconductor device, such as either an NPN or PNP transistor, in which there are both majority and minority carriers.

bipolar power supply A regulated DC power supply that can be set to provide any given voltage between positive and negative design limits, while the transition from one polarity to the other is smooth.

bipolar transistor See *transistor*.

biquad filter See *biquartic filter*.

biquartic filter An active, high-Q, bandpass filter using operational amplifiers to form an inverting amplifier and two identical integrators. Also called a *biquad filter*.

birdie A high-pitched whistle sometimes heard while tuning a radio receiver as a result of beating between two carrier frequencies that differ by approximately 10 kHz.

bistable Having two stable states.

bistable multivibrator A device that can be triggered into one of two stable output states. Also called a *flip-flop*.

bistable relay A relay that requires two pulses to complete one cycle composed of two conditions of operation. Also called a *latching, locked,* or *interlocked relay*.

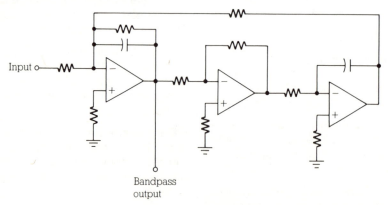

Biquartic filter

black body An ideal object that absorbs all incident radiation while radiating energy solely as a function of temperature.

black box A generalized term for an electrical circuit or system considered to be accessible only at its terminals. Frequently it is not required to know exactly what the actual circuit is.

blank To cut off the electron beam of a cathode-ray tube.

bleeder resistor A resistor connected in parallel with a DC voltage supply to improve voltage regulation by continuously drawing a fixed amount of current. It is also used to safely discharge filter capacitors when the power supply is turned off.

block diagram A diagram in which the major functions of an electronic circuit or system are indicated by simple geometric shapes, such as rectangles, and the signal paths are indicated by lines with arrowheads.

blocked impedance The impedance at the input of a transducer when the impedance of the output system is made infinite, such as by blocking or clamping off the mechanical system.

blocking (1) Applying a high negative bias to the grid (base, or gate) of a vacuum tube (transistor, or FET) to block its plate (collector, or drain) current. (2) Overloading of a receiver by an unwanted signal so that the automatic gain control reduces the response to a desired signal.

blocking capacitor See *coupling capacitor.*

blocking oscillator An oscillator circuit which behaves regeneratively for a short period of time during which the tube or transistor is driven into cutoff. It remains there until the circuit's time constant allows the bias to decay to a point where the action repeats itself. The oscillator's output is then a series of pulses. Also called a *squegging oscillator.*

blower An electric fan that supplies air for cooling electronic equipment

Bode diagram A frequency response diagram in which the gain or phase shift of an amplifier, filter, or other circuit is plotted as a function of frequency. *(See figure on page 16.)*

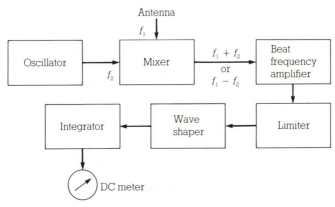

Block diagram

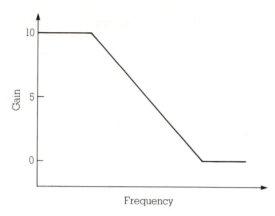

Bode plot

bolometer An instrument that measures microwave or infrared energy. It consists of a resistive element whose resistance changes when heated by the radiant energy.

bolometer bridge A four-arm bridge circuit that uses a heat-sensitive resistive element in one arm to measure RF power.

Boltzmann's constant Symbolized by the letter k, a physical constant equal to 1.380662×10^{-23} joules per degree kelvin.

bonded strain gage A type of transducer that has its strain gage elements firmly bonded to a strain-responsive member. Thermal stability and insensitivity to shock and vibration are improved with a bonded gage system.

bootstrapped amplifier A single-stage amplifier in which the output load is connected between the negative of the collector (plate) and the emitter (cathode) of a transistor (vacuum tube) while the signal voltage is applied between the base (grid) and emitter (cathode). Consequently, a change in base (grid) voltage changes the input signal with respect to ground by an amount equal to the output signal voltage.

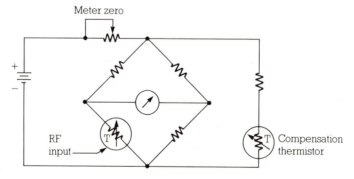

Bolometer bridge

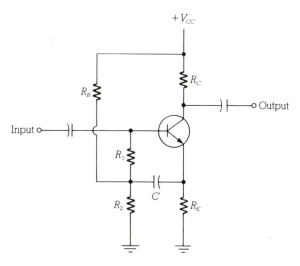

Bootstrapped amplifier

Bosanquert's law The magnetic circuit analog of Ohm's law, stating that the magnetomotive force produced in a magnetic circuit is equal to the product of the magnetic flux and the reluctance.

boxcar circuit A circuit that, during a gated interval, allows an input signal to charge a capacitor followed by a buffer amplifier so that the output is proportional to the accumulated charge stored on the capacitor.

boxcars Long pulses separated by very short intervals so that the duty cycle approaches 100 percent.

braided wire A tube or mesh of fine wires woven around a conductor or cable for shielding, or used alone as a grounding strap.

branch In an electrical network, an element, or a series connection of two or more elements, included between two junctions.

breadboard An experimental version of a circuit generally laid out on a flat board and assembled with tem-

porary connections so that circuit elements may be easily substituted or changed. The name originates from the fact that early electrical circuits were actually wired on wood breadboards.

break A fault or interruption in a circuit.

breakdown diode Also called an *avalanche diode. See Zener diode.*

breakdown impedance The small-signal impedance at a specified DC current in the breakdown region of a semiconductor diode. Also called *Zener impedance.*

breakdown region The entire region of the semiconductor diode's reverse voltage-current characteristic curve where the voltage remains constant for increasing current.

breakdown voltage The voltage measured at a specified current in the breakdown region of a semiconductor diode; also called *Zener voltage.*

break frequency See *cutoff frequency.*

bridge A circuit configuration that has four arms or branches connected in series in the shape of a parallelogram. Two opposite nodes of the bridge serve as the input terminals while the two remaining nodes are the output terminals.

bridge amplifier A differential amplifier whose inputs are connected to the outputs of a bridge circuit in place of a null detector or meter.

bridged-T network A T-network with a fourth branch element connected across the two series elements of the T from the input and output terminals.

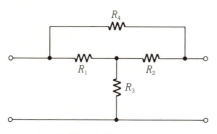

Bridged-T network

bridge rectifier A full-wave rectifier with four diodes connected in series in the form of a bridge circuit. AC voltage is applied to one pair of opposite nodes while DC voltage is obtained from the other pair of nodes.

broadband (1) Having essentially a constant or uniform frequency response over a wide range of frequencies. (2) Wideband.

broadcast band For AM broadcasting, the frequency range from 540 to 1 600 kHz, divided into channels of 10 kHz bandwidth each.

Bruce antenna See *rhombic antenna.*

bucking voltage A voltage that has exactly the opposite polarity to that of another voltage against which it acts.

buffer (1) An isolating circuit used to prevent unwanted interaction between two parts of a circuit. (2) A buffer amplifier.

buffer amplifier (1) A unity-gain amplifier having a high input impedance and a very low output impedance, used to minimize interaction between two circuits or stages. (2) An operational, amplifier voltage follower, cathode follower, emitter follower, or source follower amplifier.

bulb A lamp or light bulb. See *envelope.*

burst amplifier An amplifier in a color television receiver keyed into conduction and amplification by a horizontal pulse at the exact arrival of the 3.58 MHz color burst signal. Also called a chroma bandpass amplifier.

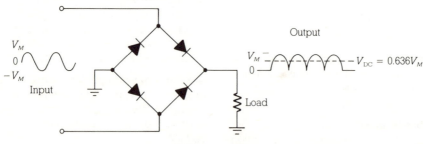

Bridge rectifier

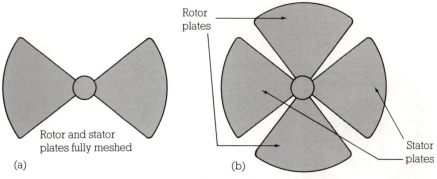

Butterfly capacitor

bus One or more conductors used as a path for the transmission of signals.

bus bar A bar or strip of copper or aluminum used as a connection between circuits, such as ground.

butterfly capacitor A variable capacitor whose stator and rotor plates are shaped like butterfly wings. Used primarily in VHF and UHF circuits, the capacitance is at a minimum when the stator and rotor plates form the four quadrants of a circle, and at a maximum when they are fully meshed.

Butterworth filter A type of active or passive filter that has essentially a flat response in the passband and a

rolloff of an integer multiple of 6 dB per octave (or 20 dB per decade), depending on the order of the filter. Also called a *maximally flat filter*.

BW Abbreviation for *bandwidth*.

BX cable Insulated wires housed inside a flexible metal tubing, which is used primarily to safely protect power line cables inside buildings and houses.

bypass A low-impedance path around part or all of a circuit.

bypass capacitor A capacitor used to provide a low-impedance path to ground at audio or radio frequencies around a component or circuit element.

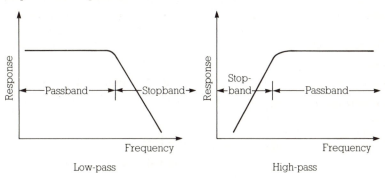

Butterworth filter

C Symbol for the *collector* lead of a transistor, *capacitor, capacitance, coulomb, Celsius (centigrade)*.

calibrate (1) To determine, either by measurement or by comparison with a known standard, the correct value of a reading on an instrument or other device. (2) To determine the required settings of a control that correspond to particular values of voltage, current, frequency, or some other parameter.

can (1) A metal shield placed around a tube, transformer, or coil to prevent electromagnetic or electrostatic interaction. (2) A metal housing or package for enclosing semiconductor devices, such as transistors and integrated circuits.

candela Abbreviated *cd*, the unit of luminous intensity.

canibalization A method by which the required repair parts are removed from one system or assembly for installation in a similar system or assembly.

capacitance (1) Abbreviated and symbolized by the letter *C*, the property that permits the storage of electrically separated charges when a potential difference exists between a conductor on opposite sides of a dielectric. Measured in farads, equal to the ratio of the electric charge that has been transferred from one electrode of a capacitor to the other and the potential difference between the electrodes. (2) Capacity.

capacitance ratio In a variable capacitor, the ratio of the maximum to minimum capacitance.

capacitance relay An electronic circuit with a relay that responds to a small change in capacitance, such as that produced by one's body near a wire or plate.

capacitive coupling The use of a capacitor to transfer energy or AC signals from one circuit to another. The DC component of a signal is blocked.

capacitivity See *dielectric constant*.

capacitor   Symbolized by the letter *C* on schematic diagrams, a device that consists of two conducting surfaces separated by a dielectric material. A capacitor stores electric energy and blocks the flow of direct current. Formerly called a *condenser.*

capacitor bank   A number of capacitors connected in series or in parallel.

capacitor-input filter   A power-supply filter in which a shunt capacitor is the first element after the rectifier element.

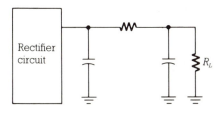

Capacitor-input filter

capacitor motor   A single-phase induction motor with the main winding arranged for direct connection to the power source together with an auxiliary winding connected in series with a capacitor.

capacity   (1) The amount of electricity that can be drawn from a battery or cell, usually expressed in ampere-hours. (2) Capacitance.

capture   In a phase-locked loop, the condition where the output frequency begins to change from its free-running frequency to that of the loop's input.

carbon resistor   A resistor composed of carbon particles which are mixed with a binder and molded into a cylindrical shape with axial leads. Also called a *composition resistor.*

carbon-zinc cell   The most common type of dry battery, using a carbon rod positive terminal in the center of the cell and a zinc can as the negative terminal with a sal ammoniac paste in between.

carborundum   A compound of carbon and silicon used to detect or rectify radio waves.

cardioid   A polar diagram in the shape of a heart, which is similar to the radiation pattern of a dipole antenna with a reflector, or a microphone. Its radius as a function of the polar angle is given by $r = a(1 + \cos \theta)$.

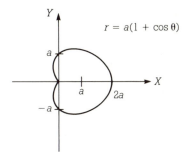

Cardioid

carrier   A transmitted radio signal where there is no modulation. Also called a *carrier wave.*

carrier frequency   The frequency generated by an unmodulated transmitted signal. See *center frequency.*

carrier level   The intensity or level of an unmodulated carrier signal at a particular point in a radio system, generally expressed in decibels in relation to a reference level.

carrier noise level   The noise level produced by unwanted variations of an RF signal without any modulation. Also called *residual modulation.*

carrier wave See *carrier.*

cascade A connection of two or more networks in series.

cascade amplifier An amplifier having two or more stages arranged in series where the output of one stage feeds the input of the next.

cascode amplifier A two-stage transistor (or vacuum tube) amplifier

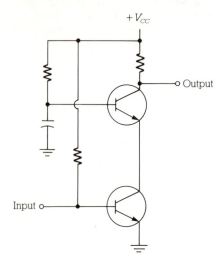

Cascode amplifier

in which the collector (plate) of the input stage feeds the emitter (cathode) of the output stage.

case temperature The temperature on the surface of a case or can of a semiconductor device.

cathode (1) Symbolized by the letter K, one of the elements in a vacuum tube. (2) The negative electrode of a battery or other electrochemical device. (3) The terminal of a semiconductor device, such as a diode, silicon-controlled rectifier, or programmable unijunction transistor.

cathode follower A common-plate vacuum tube amplifier in which the input signal is applied between the control grid and ground while the output is taken from the cathode. While characterized by heavy negative feedback, the output signal is in phase with the input, the maximum voltage gain is unity, and the input impedance is high while the output impedance is low. Used primarily as a buffer amplifier.

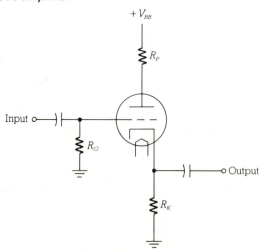

Cathode follower

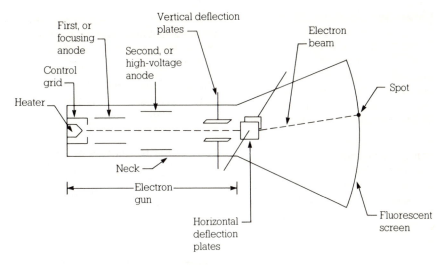

Cathode-ray tube

cathode-ray oscilloscope Abbreviated *CRO*, a test instrument that uses a cathode-ray tube to make visible on a phosphor-coated screen the instantaneous values and waveforms of electrical quantities which are rapidly changing as a function of time or other quantity. Also called an *oscilloscope* or *scope*.

cathode-ray tube Abbreviated *CRT*, an electron-beam tube in which the electrons emitted by a hot cathode are formed by an electron gun into a narrow beam that can be focused onto a phosphor-coated screen. The beam can then be varied in its position and intensity to produce a visible trace, pattern, or picture on the screen. Originally called a Braun tube, a cathode-ray tube is also called a kinescope, and, when used in a television, a picture tube.

Cauer filter An active or passive low-pass or high-pass filter whose passband is flat while the stopband has a series of decreasing peaks. The

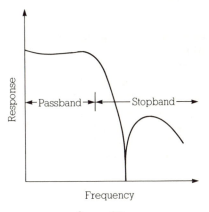

Cauer filter

rolloff is the sharpest of all the filter types of the same order. Also called an *elliptical filter*.

C-band The radio-frequency band from 3.9 to 6.2 GHz.

cd Unity symbol for *candela*.

cell A single unit that produces electricity from either chemical, radiant, or thermal energy.

Celsius temperature Abbreviated by the unit symbol *C*, developed by Anders Celsius, a temperature scale in which zero degrees is based on the triple point of water, at 0.01 degree centigrade, in place of the ice point. The centigrade temperature scale was changed to the Celsius scale in 1948.

center frequency (1) The frequency of a bandpass filter or tuned amplifier at which the output voltage is at a maximum. Also called *carrier frequency,* or *midband frequency.* (2) For a bandpass response, equal to the geometric mean of its upper and lower cutoff frequencies.

center tap Abbreviated *CT*, a terminal at the electrical midpoint of a coil or transformer.

centigrade temperature scale Abbreviated by the unit symbol *C*, a temperature scale in which the freezing point of water is zero degrees and its boiling point is 100 degrees. It is now called the Celsius temperature scale. In terms of a Fahrenheit temperature, the equivalent centigrade temperature is computed from

$$C = (°F - 32) \times \frac{5}{9}$$

channel (1) A band of radio frequencies allocated for a specific purpose, such as television. (2) A path for a signal, such as left and right stereo channels. (3) The main current path between the drain and source terminals in a field effect transistor.

channel separation (1) The electric or acoustic difference between the right and left channels in a stereo system. (2) In a multi-operational

amplifier package, the amount of the output of an unused operational amplifier section that is from crosstalk from any of the used sections.

characteristic curve A curve plotted to show the relation between two changing values, such as the collector voltage and current in a transistor.

characteristic impedance Symbolized by Z_o, the impedance that, when connected to the output of a transmission line of any length, makes the line appear infinitely long. As a result, there are no standing waves. Also called *surge impedance.*

charge (1) Symbolized by the letter *Q*, a quantity of electric energy, expressed in coulombs, stored in a capacitor, battery, elementary particle, or insulated object. Also called *electric charge.* (2) The conversion of electrical energy to chemical energy in a storage battery by direct current through the battery in the opposite direction to that of the discharge current.

charged particle An elementary particle that has either a positive or negative charge; an ion.

charger A device used to convert alternating current into a pulsating direct current for charging a storage battery.

chassis The metal frame on which circuit components are mounted.

chassis ground A ground connection to the chassis.

chatter The mechanical noise generated by the repeated opening and closing of the electrical contacts of a relay after they have been brought together.

cheater cord A type of extension power cord used to supply AC power to radio and television sets. The power is disconnected from the unit when the back cover is removed.

Chebyshev filter Also spelled *Tchebycheff* and *Tschebycheff,* an active or passive low-pass or high-pass filter that has equal amplitude ripples in the passband but a sharper rolloff than that of a Butterworth filter of the same order. Also called an *equal ripple filter,* or *constant-k filter.*

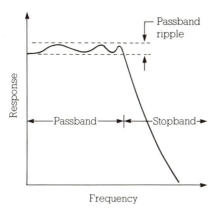

Chebyshev filter

chi (χ) The twenty-second letter of the Greek alphabet, used as a symbol for *susceptibility.*

choke An inductor used to present a high impedance to frequencies above a given frequency without appreciably limiting the flow of DC current.

chopper A device or circuit for interrupting the flow of current, light, or other energy at a regular rate.

chopper amplifier A carrier amplifier in which a DC input is filtered by a low-pass filter, converted into a square wave AC signal by a chopper, and then amplified. After amplifica-

tion, the square wave is rectified by chopper action to obtain a DC output.

chroma The quality, excluding brightness, that identifies any particular color.

circuit An electrical network in which there exists one or more paths through which current can flow.

circuit breaker A switching device that automatically interrupts an electric current from its load when the current exceeds a given value for too long a time.

circuit diagram See *schematic.*

clamper See *DC restorer.*

clamping circuit A diode circuit that reestablishes the DC level of a waveform.

clamping diode A diode that clamps a voltage at some point in a circuit.

clapper A relay armature that is hinged or pivoted.

Clapp oscillator A Colpitts oscillator that is modified so that it is tuned by a series LC circuit, providing improved frequency stability. *(See figure on page 26.)*

class-A amplifier A linear amplifier in which the operating point is chosen so that output current flows at all times.

class-AB amplifier A linear amplifier in which the operating point is chosen so that output current flows for more than half but less than the entire input cycle.

class-B amplifier A linear amplifier in which the operating point is biased at cutoff so that output current flows for half of each input cycle.

25

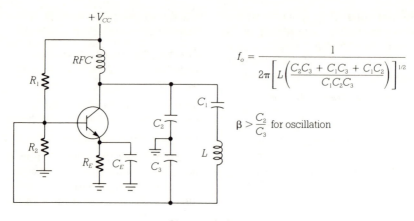

$$f_o = \cfrac{1}{2\pi\left[L\left(\cfrac{C_2C_3 + C_1C_3 + C_1C_2}{C_1C_2C_3}\right)\right]^{1/2}}$$

$\beta > \dfrac{C_2}{C_3}$ for oscillation

Clapp oscillator

clip To limit in value.

clipper A diode circuit that limits the peak value of a signal which otherwise would result in distortion in an amplifier.

closed loop A signal path in which the output of a control system is fed back to the system's input.

closed-loop bandwidth The frequency at which the closed-loop gain decreases by 3 dB from its midband or DC value.

closed-loop gain The gain of an amplifier with an external negative feedback path.

closed-loop voltage gain The voltage gain of an amplifier with feedback.

CM Abbreviation for *common mode*.

CMOS Abbreviation for *complementary metal oxide semiconductor*.

CMR Abbreviation for *common-mode rejection*.

CMRR Abbreviation for *common-mode rejection ratio*.

C-network A network of three impedance branches connected in series in the shape of the letter *C*.

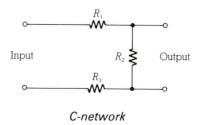

C-network

coaxial cable A transmission line in which one conductor is completely surrounded by the other in the form of a metal sheath or braid. The two conductors have a common longitudinal axis and are separated by a dielectric. Such a line produces no external field and cannot be affected by an external field. Also called *coaxial line, coaxial transmission line, concentric line,* or *concentric transmission line.*

coaxial line See *coaxiable cable.*

coaxial transmission line See *coaxial cable.*

coil   Designated by the letter L on schematic diagrams, a number of turns of wire in the form of an inductor.

cold junction   The junction of thermocouple wires with conductors leading to the measuring instrument. This junction is either at the temperature of freezing water or at room temperature.

collector   Abbreviated C, one of the three terminals of a bipolar transistor into which majority carriers flow from the base region under the influence of a reverse bias across the two regions.

Colpitts oscillator   An oscillator in which a parallel-tuned LC tank circuit has two voltage dividing capacitors.

common-base amplifier   A transistor amplifier in which the base terminal is common to both the input and output circuits. The input signal is applied to the emitter and the output is taken from the collector, which is in phase with the input. Also called a *grounded-base amplifier.*

common-collector amplifier   A transistor amplifier in which the collector terminal is common to both the input and output circuits. The input signal is applied to the base and the output is taken from the emitter, which is in phase with the input. Also called an *emitter follower amplifier.*

common-drain amplifier   A field effect transistor amplifier in which the drain terminal is common to both the input and output circuits. The input signal is applied to the gate and the output is taken from the source, which is in phase with the input. Also called a *source follower amplifier.*

common-emitter amplifier.   A transistor amplifier in which the emitter terminal is common to both the input and output circuits. The input signal is applied to the base and the output is taken from the collector, which is 180 degrees out of the phase with the input. Also called a *grounded-emitter amplifier.*

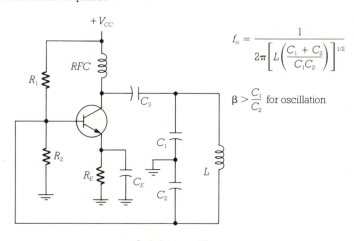

$$f_o = \frac{1}{2\pi\left[L\left(\dfrac{C_1 + C_2}{C_1 C_2}\right)\right]^{1/2}}$$

$\beta > \dfrac{C_1}{C_2}$ for oscillation

Colpitts oscillator

27

common-gate amplifier. A field effect transistor amplifier in which the gate terminal is common to both the input and output circuits. The input signal is applied to the source and the output is taken from the drain, which is in phase with the input. Also called a *grounded-gate amplifier.*

common mode Abbreviated *CM*; in a differential amplifier, signals that are identical in phase and amplitude at both inputs simultaneously.

common-mode gain In a differential amplifier, the ratio of the common-mode output voltage to the common-mode input voltage.

common-mode input The input signal applied equally and simultaneously to both inputs of a differential amplifier.

common-mode rejection Abbreviated *CMR*, the ability of a differential amplifier to cancel a common-mode signal. Expressed in decibels, it is related to the common mode rejection ratio so that CMR = 20 log(CMRR).

common-mode rejection ratio Abbreviated *CMRR*, the ability of a differential amplifier to cancel a common-mode signal. It is equal to the ratio of the common-mode input voltage to the generated output voltage.

common-source amplifier A field effect transistor amplifier in which the source terminal is common to both the input and output circuits. The input signal is applied to the gate and the output is taken from the collector, which is 180 degrees out of phase with the input. Also called a *grounded-source amplifier.*

commutating angle In an inverter or rectifier circuit, the fraction of an AC cycle, expressed in electrical degrees, during which current is commutated from one rectifying element to another.

commutation The transfer of current from one anode to another in a rectifier circuit.

commutator A circular arrangement of copper bars insulated from each other and the rotor of a DC motor or generator on which they are mounted, with current-carrying brushes bearing down on the exposed surfaces in order to provide the transfer of direct current to the armature coils in sequence.

compander An electronic circuit that combines the functions of a compressor and a complementary expander. The compressor reduces the volume range of the audio input signal while the expander improves the signal-to-noise ratio.

comparator An active circuit that compares an input voltage to a predetermined DC voltage reference.

comparison gate See *exclusive-OR gate.*

compensated amplifier A broadband amplifier in which the frequency range is extended by a choice of circuit constants.

compensation Modification of the amplitude frequency response of an amplifier either to increase the bandwidth or to make the response nearly uniform over the existing bandwidth. Also called *frequency compensation.*

complementary MOS Abbreviated *CMOS*, a combination of n- and p-channel enhancement mode metal oxide semiconductor field effect transistors on a single silicon chip.

complementary transistors Two transistors whose characteristics and ratings are similar but opposite in their construction, such as NPN and PNP bipolar transistors, and n- and p-channel field effect transistors.

component Any electric device that has distinct electrical character-istics, and terminals that may be con-nected to other components to form a circuit. Also called an *element*.

composite picture signal The complete picture signal as it leaves the television transmitter. The signal consists of picture data, horizontal and vertical blanking pulses, syn-chronizing pulses for monochrome (black and white) pictures, or color subcarrier and color burst for color pictures. Also called *composite video signal*.

composite video signal See *composite picture signal*.

composition resistor See *carbon resistor*.

compressor An audio amplifier whose gain is automatically adjusted to the signal level in such a way that weak signals are amplified with more gain than strong signals, as in the action of a logarithmic amplifier.

concentric line See *coaxial cable*.

concentric transmission line See *coaxial cable*.

condenser See *capacitor*.

conductance (1) Symbolized by the letter G, a measure of the ability to pass electrical current, and equal to the reciprocal or resistance. (2) The real part of admittance.

conductivity Symbolized by the Greek letter σ, the reciprocal of resistivity.

conductor A wire, cable, or other medium suitable for carrying electric current.

conjugate bridge A bridge circuit in which the detector and voltage source elements are interchanged, compared with the normal bridge of the same type.

conjugate impedances Imped-ances whose real parts are equal, but whose imaginary parts are equal in magnitude but opposite in polarity. If one complex impedance is $z = x + jy$, then $z^* = x - jy$ is the conjugate impedance.

connection A direct wire path between two points in a circuit.

constant-*k* filter See *Chebyshev filter*.

constant-*k* network A network in which the product of the series and shunt impedances is a constant, and independent of frequency. *(See figure on page 30.)*

contact A conducting part of a relay, connector, or switch, which can be connected with another such part to make or break a circuit.

contact bounce The uncontrolled making and breaking of contact of a switch or relay many times when the initial contact is made.

contact resistance The resistance measured across a pair of closed contacts of a switch or relay.

continuity The presence of a com-plete path for current flow.

continuous-duty rating The rating that defines the load that can be car-ried for an infinite amount of time without exceeding a specified rise in temperature.

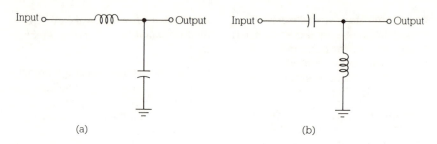

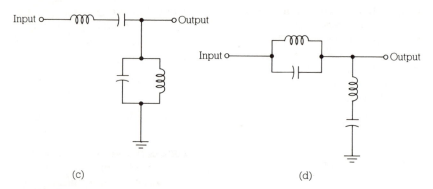

*Constant-*k *network filters: (a) Low-pass; (b) High-pass; (c) Bandpass; (d) Notch*

control grid   The electrode in a vacuum tube whose potential controls the current flow between the plate and cathode. Also called the first grid or G_1.

conversion gain   The ratio expressed in decibels of output signal power to input signal power for a signal converter or mixer. Also called conversion gain ratio.

copper loss   Power loss in a winding of a coil due to current flow through the resistance of the copper conductors. Also called I^2R loss.

core   See *magnetic core*

core loss   The loss of energy in the magnetic core of an inductor or trans-

former as a result of eddy currents and hysteresis. Also called *iron loss.*

corner frequency   See *cutoff frequency.*

coulomb   Abbreviated C, the unit of electric charge equal to the charge that passes through a given cross section of a conductor when a DC current of 1 ampere is flowing for 1 second. Equal to the charge carried by 6.281×10^{18} electrons.

Coulomb's law   The attraction of two electric charges is proportional to the product of the magnitude of the two charges, and inversely proportional to the square of the distance between them. Also called the *law of electrostatic attraction.*

30

coupling   The relationship between two systems or networks such that energy is transferred from one to the other while each influences the behavior of the other.

coupling capacitor   A capacitor that blocks the flow of DC signals while allowing AC signals to pass between two networks. Also called a *blocking capacitor.*

coupling coefficient   Symbolized by the letter *k* and by the Greek letter κ, a rating between 0 and 1 (maximum coupling) that quantifies the degree of coupling between two circuits.

crest   See *peak.*

crest factor   The ratio of the peak value to the effective, or root mean square value of a periodic waveform. For a sinewave, the crest factor is 1.414. Also called *peak factor.*

crest value   See *peak value.*

critical coupling   (1) The degree of coupling between two resonant systems that results in a maximally flat response around the resonant frequency. (2) Optimum coupling.

critical damping   The degree of damping required to give the most rapid transient response without overshoot or oscillation.

critical frequency   See *cutoff frequency.*

critical inductance   The minimum inductance required of a choke input of a full-wave rectifier to prevent the current through the choke from decreasing to zero during any part of the cycle.

CRO   Abbreviation for *cathode-ray oscilloscope.*

cross coupling   Undesired coupling between circuits.

crossover   A point at which two insulated wires or conductors cross.

crossover distortion   The distortion produced in a class-B push-pull amplifier when the vacuum tubes or transistors are biased exactly at cutoff. The distortion occurs in the region where the output voltage is zero.

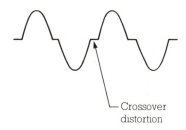

Crossover distortion

Crossover distortion

crossover frequency   The frequency at which a crossover network delivers the same power to the upper and lower frequency channels when both are terminated with specified loads. Also called *transition frequency.*

crossover network   A network that divides an audio frequency signal into two or more bands, each of which is fed to a load.

cross talk (or crosstalk)   Any unwanted signal transfer between two channels or devices. The measure of separation is known as channel separation.

crowbar   A low-resistance shunt quickly placed across the output terminals of a power supply as a protective measure.

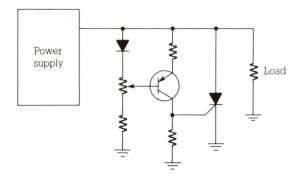

Crowbar circuit

crowbar circuit A circuit that monitors the output of a power supply and quickly places a low-resistance shunt, called a crowbar, across the output terminals whenever a preset voltage limit is exceeded. This provides protection until slower fuses or circuit breakers can respond.

CRT Abbreviation for *cathode-ray tube.*

crystal Abbreviated *xtal,* a natural or synthetic piezoelectric or semiconductor material whose atomic structure has a specific geometric arrangement.

crystal lattice filter A crystal filter that has two matched pairs of series crystals and a higher-frequency matched pair of shunt crystals.

crystal oscillator An oscillator whose output frequency is controlled by the resonant frequency of a piezoelectric crystal.

crystal oven A temperature-controlled enclosure in which the crystal element of a crystal-controlled circuit is placed to stabilize its temperature, thereby minimizing frequency drift.

CT (or C.T.) Abbreviation for *center tap.*

curl The cross product of the vector operator ∇ (del) and another vector, **A**. In Cartesian coordinates:

$$\text{curl } \mathbf{A} = \nabla \times \mathbf{A}$$

$$= \mathbf{i}\left(\frac{\partial V_z}{\partial y} - \frac{\partial V_y}{\partial z}\right)$$

$$+ \mathbf{j}\left(\frac{\partial V_x}{\partial z} - \frac{\partial V_z}{\partial x}\right)$$

$$+ \mathbf{k}\left(\frac{\partial V_y}{\partial x} - \frac{\partial V_x}{\partial y}\right)$$

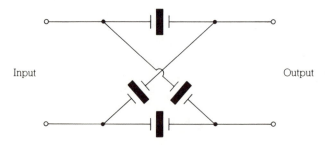

Input   Output

Crystal lattice filter

current symbolized by the letter *I*, the time rate of change in which charge is transferred, $I = dQ/dt$, as measured in amperes.

current drain The current taken from a voltage source by a load.

current foldback A regulated power supply circuit feature that protects against overloading and short circuits, rather than maintaining a constant-current operation. Also called *current limiting.*

current gain Abbreviated A_I, the ratio of the output current of a network to the input current. Expressed in decibels (dB), the current gain is

$$A_I = 20 \log \frac{I_{out}}{I_{in}}$$

current generator A two-terminal circuit element whose output current is independent of the voltage between its terminals.

current limiting See *current foldback.*

current relay A relay that is actuated by a specified value of current through its coil rather than by a specified voltage across it.

current transfer ratio See *beta.*

curve tracer An oscillographic instrument that automatically displays the characteristic curve of an electronic component.

cutoff (1) In a vacuum tube or transistor, the condition in which there is no plate or collector current flow. (2) The bias voltage required to produce cutoff.

cutoff frequency (1) The frequency at which the voltage of a frequency-sensitive network decreases to 0.707 times its midband value. Also called *frequency cutoff, break frequency, critical frequency, corner frequency,* and *3-dB frequency.* (2) In terms of decibels, the frequency at which the voltage gain decreases by 3 dB from its midband value.

cycle One complete sequence of values of a periodic quantity, which includes the rise to a maximum in one direction, a return to zero, a decrease to a minimum in the other direction, and another return to zero.

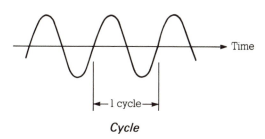

Cycle

d Prefix symbol for *deci*, 10^{-1}.

D (1) Mathematical symbol for *flux density, dissipation factor*. (2) Abbreviation for *drain*. (3) Letter designation for *diode* on schematic diagrams.

DA Abbreviation for *distribution amplifier*.

D/A Abbreviation for *digital-to-analog*.

DAC Acronym for *digital-to-analog converter*.

damping Any action that extracts energy from a vibrating system to suppress the vibration or oscillation.

damping coefficient Symbolized by the Greek letter ζ, the ratio of actual damping to critical damping.

damping constant The Napierian logarithm of the damping ratio. Also called *logarithmic decrement*.

damping factor For an underdamped system, the ratio of the logarithmic decrement or damping constant to the period of one cycle.

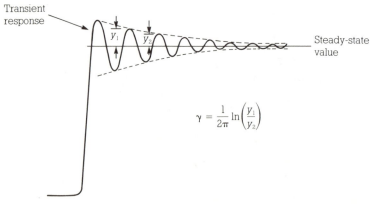

Damping constant

damping ratio (1) The ratio of two successive peak values of the same polarity of an exponentially damped oscillation. (2) The inverse Naperian logarithm of the damping constant.

daraf Farad spelled backwards, a unit of elastance equal to the reciprocal of capacitance.

dark current The current flowing through a photoelectric device in the absence of light.

dark resistance The resistance of a photoelectric device in total darkness.

Darlington connection A connection of two like transistors to act as a single transistor so that the overall current gain is the product of the two individual current gains. Both collectors are connected together and the emitter of one transistor is connected to the base of the other. Also called a Darlington amplifier or a double emitter follower.

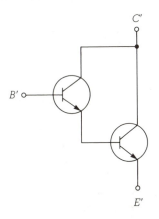

Darlington connection

D'Arsonval movement A permanent-magnet moving coil instrument.

dB Unit symbol abbreviation for *decibel.*

dBm Unit symbol abbreviation for decibels above 1 milliwatt, a ratio of power in terms of a power level referenced to 1 milliwatt.

DC (or dc) (1) Abbreviation for *direct current.* (2) Zero frequency.

DC amplifier Abbreviation for direct current amplifier, an amplifier using resistive or direct coupling capable of amplifying DC signals.

DC component The average value of a signal.

DC coupling The coupling of one circuit to another whereby only the DC component of an electrical signal is passed through.

DC current Abbreviation for *direct current* current; mathematically symbolized by I. Although the term is redundant and awkward, it has been frequently used to differentiate the current from DC voltage.

DC/DC converter An electric circuit that converts a DC voltage to a different value.

DC generator A rotating machine that converts mechanical power into DC electrical power.

DC motor A motor that operates from a DC voltage source and converts electrical energy into mechanical energy.

DC offset A DC level that may be added to the input signal of an amplifier or other circuit.

DC power supply A power supply that provides one or more DC voltages.

DC relay A relay whose coil responds to a DC voltage.

DC resistance The resistance between two points in a DC circuit.

DC restorer A diode-type circuit used in television receivers to restore the DC component of the video signal after amplification. Also called a *clamper* or *restorer*.

DC voltage Abbreviation for *direct current* voltage; mathematically symbolized by *V*. Although awkward, the term has been frequently used to distinguish the voltage from DC current.

dead De-energized; not containing or carrying electricity.

decade (1) A multiple of 10. (2) Any interval between two quantities having a ratio of 10:1.

decade box A circuit made entirely of precision capacitors, inductors, or resistors whose output can be adjusted to vary in submultiples and multiples of 10.

decay time The time taken for a quantity, such as voltage, persistance, etc., to decrease to a given fraction of its initial value.

deci Abbreviated *d*, prefix for 10^{-1}.

decibel Abbreviated *dB*, a logarithmic unit, equal to 0.1 bel, used to express the ratio of one signal level to another. In terms of power levels,

$$dB = 10 \log \frac{P_2}{P_1}$$

In terms of voltage levels,

$$dB = 20 \log \frac{V_2}{V_1}$$

decouple To arrange two or more circuits so that coupling between them is eliminated.

decoupling network A network of capacitors, resistors, or inductors (chokes) placed into leads of one or more circuits to prevent unwanted coupling of signals.

de-emphasis The process of reducing the relative intensity of higher audio frequencies before amplification. Also called postemphasis or postequalization.

de-emphasis network A filter network, either active or passive, that restores pre-emphasized signals to their original form.

degauss (1) To demagnetize. (2) To remove or erase information from a magnetic tape, disk, drum, or core.

degeneration See *negative feedback*.

degree (1) A measure of an angle, equal to $2\pi/360$ radians. (2) A measure of temperature, defined in numerous ways depending on the system of measurement. (3) In mathematics, the value of the highest power of a polynomial equation, or the highest order derivative in a differential equation.

del The vector operator symbolized by ∇, defined in Cartesian coordinates as

$$\nabla = \mathbf{i}\frac{\partial}{\partial x} + \mathbf{j}\frac{\partial}{\partial y} + \mathbf{k}\frac{\partial}{\partial z}$$

where **i**, **j**, and **k** are unit vectors along the *X*, *Y*, and *Z* axes.

delay distortion A form of phase distortion in which the rate of phase shift with frequency is constant over a required frequency range. Also called envelope delay distortion.

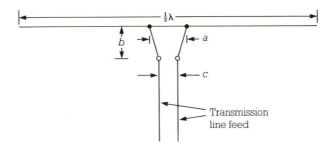

Delta match antenna

delay line   An electronic circuit or transmission line used to produce a time delay of a signal.

delta-match antenna   A single-wire, half-wavelength antenna fed by an open-wire transmission line connected in the shape of a Y.

delta network   A two-port, three-branch network in the form of the Greek letter delta (Δ) or the Greek letter pi (Π). Also called a π (or pi-) network.

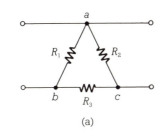

(a)

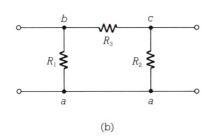

(b)

Delta network:
(a) Standard arrangement;
(b) Equivalent pi-network

demodulate   To remove the carrier component from a modulated RF signal to a form that can be either heard or displayed. To detect.

demultiplex   To separate two or more signals transmitted over a single channel.

depth of null   See *null depth.*

derate   To reduce the current, power, or voltage rating of a device either to improve its reliability or to permit operation at higher ambient temperature.

derating factor   The factor by which a component rating is reduced to provide an additional safety margin when the component is subjected to extreme conditions for which the normal rating does not apply.

detect   See *demodulate.*

deviation   See *frequency deviation.*

deviation ratio   In frequency modulation, the ratio of the maximum change in carrier frequency to the highest modulating frequency.

device   A single, discrete electronic part, or a group of parts taken collectively to form a circuit.

37

device under test Abbreviated *DUT*, any device that is connected to a special circuit for checking performance characteristics.

diac Acronym for diode AC switch, a bidirectional, solid-state diode that has a symmetrical switching mode. Also called a *trigger diode*.

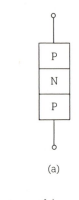

(a)

(b)

(c)

Diac: (a) Internal structure;
(b) V-I characteristic;
(c) Symbol

dielectric (1) The insulating medium between the parallel plates of a capacitor. (2) A medium capable of recovering, as electrical energy, all or part of the energy required to establish an electric field.

dielectric constant Symbolized by the Greek letter ϵ, and expressed in henries per meter, a dimensionless quantity equal to the ratio of the capacitance of a capacitor with a given dielectric to an identical capacitor with air for its dielectric. Also called *permittivity, capacitivity*.

dielectric strength The maximum voltage, expressed in terms of volts per mil, a dielectric can withstand without rupturing.

difference amplifier See *differential amplifier*.

difference frequency In a frequency converter or mixer, the output frequency equal to the difference of its two input frequencies.

differential amplifier An amplifier having high common-mode rejection and whose output is proportional to the difference of its two input signals. Also called a *difference amplifier*.

differential gain In a differential amplifier, the ratio of the output voltage to the differential input voltage.

differential transformer See *linear variable differential transformer*.

differential winding A coil winding arranged so that its magnetic field opposes that of a nearby coil.

differentiator (1) An electrical network, either active or passive, whose output is proportional to the derivative, or time rate of change of its input signal. (2) A high-pass filter.

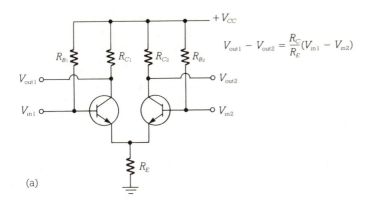

$$V_{out1} - V_{out2} = \frac{R_C}{R_E}(V_{in1} - V_{in2})$$

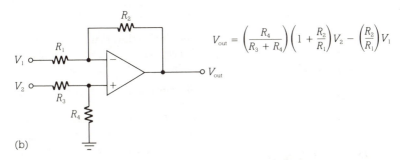

$$V_{out} = \left(\frac{R_4}{R_3 + R_4}\right)\left(1 + \frac{R_2}{R_1}\right)V_2 - \left(\frac{R_2}{R_1}\right)V_1$$

Differential amplifier: (a) Transistor version; (b) Operational amplifier version

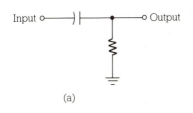

(a)

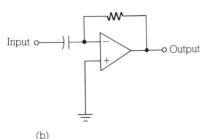

(b)

Differentiator: (a) Passive; (b) Active

digital-to-analog converter Abbreviated *D/A converter* or *DAC*, an electronic circuit that changes digital data in binary form into an equivalent output voltage.

digital voltmeter Abbreviated *DVM*, a voltmeter whose indicating device is a digital display rather than an analog meter.

DIN Acronym for Deutsche Industrie Nommenausschus, a German 5-pin plug/socket standard.

39

diode Symbolized by the letter D on schematic diagrams, a two-terminal active device in which the current through it is a nonlinear function of the voltage across it. Current flows through it when its anode terminal is more positive than its cathode (forward-biased); otherwise it behaves as an open circuit.

Anode   Cathode

(a)

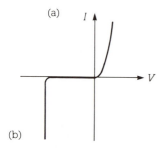

(b)

Diode: (a) Symbol;
(b) V-I characteristic

dip meter An absorption wave-meter.

dipole (1) Two particles having opposite charges or magnetic polarities separated by a small distance. (2) A dipole antenna.

dipole antenna A center-fed, straight radiator whose maximum radiation is in a plane normal to its axis. Also called a *dipole*.

direct coupling The coupling of two circuits without capacitors or inductors so that both AC and DC signals can flow through the coupling path.

direct current Abbreviated *DC*, a constant-value electric current that flows only in one direction.

disc capacitor A small, disc-shaped capacitor.

discrimination ratio The ratio of the bandwidth of a filter's passband to the width of its stopband.

discriminator A circuit whose output voltage magnitude and polarity are a function of the frequency difference between the input signal and that of another signal or reference frequency.

dissipate To lose energy, generally by conversion to heat.

dissipation factor Symbolized by the letter D, the reciprocal of the Q, or storage factor of an inductor or capacitor.

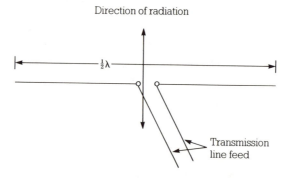

Dipole antenna

distortion Any undesired change in the waveform of a signal, especially in shape or phase.

distributed A parameter, such as resistance, capacitance, or inductance, spread out over an electrically significant length or area.

distribution amplifier (1) Abbreviated *DA*, an RF power amplifier used to feed television or radio signals to a number of receivers. (2) An AF power amplifier, used to feed a speech or music distribution system, which has a low output impedance so that changes in the load do not affect the output voltage.

div Abbreviation for *divergence,* a vector operation.

divergence (1) The spreading of a cathode-ray stream due to the repulsion of like charges (electrons). (2) A scalar quantity, abbreviated *div*, equal to the dot product of the del operator (∇) and a vector **A**, so that in Cartesian coordinates:

$$\text{div } \mathbf{A} = \nabla \cdot \mathbf{A}$$
$$= \frac{\partial A_x}{\partial x} + \frac{\partial A_y}{\partial y} + \frac{\partial A_z}{\partial z}$$

Doppler, Christian Johann An Austrian physicist and mathematician (1803–1853) who investigated the effect of the apparent frequency resulting from the motion of a source or a stationary observer. See *Doppler effect.*

Doppler effect Named after the Austrian mathematician Christian Doppler, the change in observed frequency of a wave due to the relative motion of the source to the observer. The observed frequency increases as the distance decreases, and vice versa.

dot product An operation on two vectors which results in a scalar quantity. For two vectors **A** and **B** separated by an angle θ, the dot product in Cartesian coordinates is:

$$\mathbf{A} \cdot \mathbf{B} = |\mathbf{A}| \, |\mathbf{B}| \cos \theta$$
$$= A_x B_x + A_y B_y + A_z B_z$$

Also called the *scalar product.*

double bridge See *Kelvin bridge.*

double emitter follower See *Darlington connection.*

double-tuned circuit A circuit that is resonant to two adjacent frequencies so that there are two approximately equal values of peak response with a dip in between.

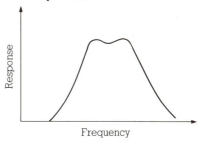

Double-tuned circuit response.

down lead The transmission line that connects an antenna to a transmitter or receiver.

DPDT Abbreviation for double pole, double throw. A part of a switch having two separate wipers, each with the ability to switch an electrical current to one of two paths.

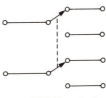

DPDT switch

41

DPST Abbreviation for double pole, single throw. A part of a switch having two separate wipers, each with the ability to switch an electrical current on or off.

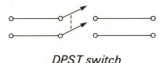

DPST switch

drain Abbreviated *D*, the lead of a field effect transistor that corresponds to the collector of a bipolar transistor or the plate (anode) of a vacuum tube.

drift A slow change in an electrical quantity, such as voltage or frequency, over a period of time.

drive See *excitation.*

driver (1) An electronic circuit which supplies an input to another circuit. (2) The amplifier stage before the power amplifier stage.

driving-point impedance See *input impedance.*

droop Expressed as a percentage, the decrease in the mean amplitude of a pulse over a given time following the initial occurrence of maximum amplitude.

dropout current The maximum current at which a relay or other electromagnetic device will release to its de-energized position.

dropout voltage The maximum voltage at which a relay or other electromagnetic device will release to its de-energized position.

dry cell A voltage-generating cell that has an immobilized electrolyte in the form of a paste surrounding a carbon anode and a zinc cathode.

dual-conversion receiver A superheterodyne receiver having two mixer stages.

dummy load An artificial, noninductive device used as a load at the end of a transmission line or waveguide to convert transmitted energy into dissipated heat. Consequently, no energy is radiated or reflected back towards the source.

DUT Abbreviation for *device under test.*

duty cycle Usually expressed as a percentage, equal to the ratio of the time a device is conducting to the total cycle time. In digital systems, the ratio of the time a pulse is at logic 1 to the total cycle.

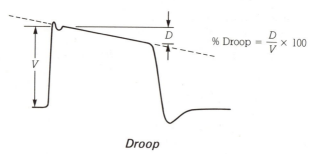

$$\% \text{ Droop} = \frac{D}{V} \times 100$$

Droop

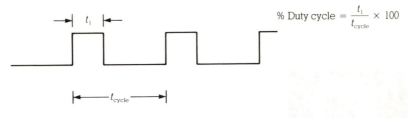

$$\text{\% Duty cycle} = \frac{t_1}{t_{cycle}} \times 100$$

Duty cycle

DVM   Abbreviation for *digital volt-meter.*

dyn   Abbreviation for *dyne.*

dynamic check   A performance check made under specific operating conditions.

dynamic range   Usually expressed in decibels, the ratio of a system or component's maximum specified signal level capability to its noise level.

dynamo   See *AC generator, alternator.*

dynamometer   An electric motor or generator that measures the torque of a rotating shaft.

dynamotor   A rotating electric machine having two or more windings on a single armature containing a commutator for DC operation and slip rings for AC operation. When one type of power is fed for motor operation, the other type of power is delivered by the generator.

dyne   Abbreviated *dyn*, the unit of force equal to the force that will cause a mass of 1 gram to accelerate at 1 centimeter per second per second. Equal to 10^{-5} newtons.

e   Symbol for the base of the *natural,* or *Napierian logarithm,* whose value is approximately 2.71828.

E   (1) Mathematical symbol for *power, permeance, electric field strength, voltage* (potential). (2) Abbreviation for *emitter.* (3) Prefix symbol for *exa* (10^{18}).

earth   British term for *ground.*

eddy current   A circulating current induced in a conducting material by a varying magnetic field. Also called *Foucault current.*

effective current   (1) The value of alternating current that will give the same heating effect as the corresponding value of direct current. For a sinewave, the effective value is 0.707 times the peak current value. (2) Root-mean square value.

effective height   The height of the center of radiation of a transmitting antenna above ground level.

effective radiated power   Abbreviated *ERP,* the product of the power input to an antenna and its power gain.

effective value   See *root-mean square.*

efficiency   Symbolized by the Greek letter η, the ratio of an output quantity to a similar input quantity, such as power.

EHF (or ehf)   Abbreviation for *extremely high frequency.*

elastance   Symbolized by the letter *S,* the reciprocal of capacitance, as measured in reciprocal farads or darafs.

electric   Of, associated with, related to, arising from, containing, or operated from electricity.

electrical angle   Symbolized usually by the Greek letters θ or φ, the angle usually expressed in electrical degrees, associated with part of an AC cycle.

electrical degree   A unit of measure equal to 1/360 cycle of a periodic quantity.

electrical length The length of a conductor or transmission line, expressed in wavelengths, radians, or degrees. The distance in wavelengths is multiplied by 2π to obtain the length in radians, or by 360 to express the length in degrees.

electric charge Symbolized by the letter Q and measured in coulombs, the electric energy stored on the surface of an object. See *charge.*

electric charge density See *volume density.*

electric circuit A continuous path over which an electric current can flow.

electric current density Symbolized by the letter J, equal to the current per unit area.

electric eye See *photocell.*

electric field (1) The region around an electrically charged body in which other charged bodies are acted upon by either an attracting or repelling force. (2) The electric component of an electromagnetic field.

electric field strength Symbolized by the letter E and expressed in volts per meter, the magnitude of the force on a stationary positive charge per unit charge at a point in an electric field.

electric flux Symbolized by the Greek letter ψ, an imaginary line, each segment of which represents the direction of the electric field at that point.

electricity The phenomenon of certain particles to possess a force, either static or dynamic, that is neither gravitational nor nuclear.

electric generator A machine that transforms mechanical power into electrical power.

electric potential A measure of the work required to bring a unit of positive charge from one infinite distance to another.

electrification (1) The process of establishing an excess of positive or negative charges in a material. (2) The process of applying a voltage to a component or device.

electrode A terminal through which current flows in passing between the conducting wires of a circuit and another conductor.

electrodynamic Pertaining to electric charges in motion.

electrolyte A substance that, when molten or dissolved in liquid, dissociates into ions that act as carriers of current.

electrolytic capacitor A polarized capacitor whose plates are separated by a dielectric film of oxide. Capacitors of this type generally have values greater than 1 μF.

electromagnet A magnet consisting of a coil wound around a magnetic core, such as soft iron or steel. When current flows through the coil, the core is strongly magnetized. It is used for attracting a movable external object, such as the armature of a relay or the core of a solenoid.

electromagnetic Abbreviated *EM* or *E-M*, pertaining to the combined effect of the electric and magnetic fields of radiation, such as radio waves, or to the movement of electrons or charged particles through conductors or free space.

45

electromagnetic coupling Coupling between circuits or conductors that are mutually affected by the same electromagnetic field.

electromagnetic energy Energy associated with electromagnetic radiation, such as radio, heat, light waves, and X-rays.

electromagnetic field The field associated with electromagnetic radiation, consisting of a moving electric field and a moving magnetic field perpendicular to each other and to the direction of propagation.

electromagnetic induction The generation of a voltage in a coil by a change in the number of magnetic lines of force passing through the coil.

electromagnetic interference Abbreviated *EMI*, any electromagnetic disturbance that interferes with the performance of electronic equipment.

electromagnetic radiation Radiation associated with periodically varying electric and magnetic fields traveling at the speed of light.

electromagnetic spectrum The entire range of frequencies or wavelengths of electromagnetic radiation, extending from approximately VLF (very low frequency, 10–30 kHz) to radiation produced by cosmic rays (10^{23} Hz, or 10^{-12} cm).

electrometer A high-impedance instrument that measures voltage without drawing appreciable current.

electromotive force (1) Abbreviated *EMF*, the force that tends to produce an electric current in a circuit. (2) Voltage.

electron An atomic particle that orbits the nucleus of an atom and that has a negative charge.

electron flow (1) A current produced by the movement of electrons toward a positive terminal. (2) The direction opposite that of current flow.

electronic Pertaining to electron devices, or to circuits and systems utilizing electron devices, including electron and vacuum tubes as well as solid-state devices.

electron tube An electron device in which conduction of electricity is provided by electrons moving through a gaseous or vacuum medium within a glass envelope. Also called *vacuum tube, radio tube, tube,* and *valve* (British).

electronvolt Abbreviated *eV*, a unit of energy equal to that acquired by an electron when it passes through a potential difference of 1 volt in a vacuum. Equal to 1.60219×10^{-12} erg.

electroscope An instrument for the detection of an electric charge by means of the mechanical forces exerted between electrically charged bodies. In its simplest form, two narrow strips of gold leaf are suspended in a glass jar and are spread apart when charged. The angle between the strips is proportional to the charge.

electrostatic Pertaining to electricity at rest.

electrostatic shield A grounded metal screen, sheet, or enclosure placed around a device or between two devices to prevent electric fields from passing through.

element See *component.*

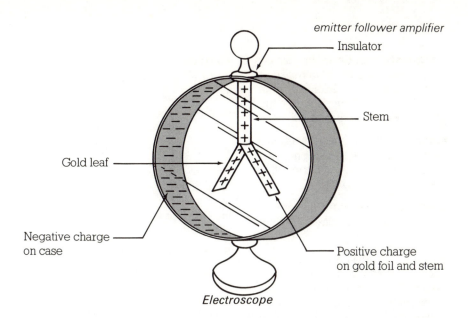

Insulator

Stem

Gold leaf

Negative charge
on case

Positive charge
on gold foil and stem

Electroscope

ELF (or elf) Abbreviation for
extremely low frequency.

elliptical filter See *Cauer filter.*

EM (or E-M) Abbreviation for
electromagnetic.

EMF (or emf) Abbreviation for
electromotive force.

EMI (or emi) Abbreviation for *electromagnetic interference.*

emitter (1) Abbreviated *E*, one of
three terminals of a bipolar transistor,
in whose region majority carriers
originate and are injected into the

base, where they either diffuse into
the collector or undergo recombina-
tion within the base. (2) In a unijunction transistor, the terminal between
the two base terminals.

emitter follower amplifier A bipolar transistor circuit whose maximum
voltage gain is unity and whose output signal is in phase with the input
signal. Used primarily as a buffer, it
exhibits a high input impedance and
a low output impedance. Also called
a *grounded-,* or *common-collector
amplifier.*

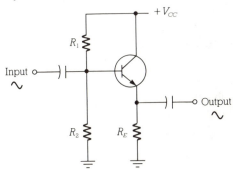

Emitter follower

emitter stabilization   The use of a Zener diode or swamping resistor to maintain essentially a constant emitter voltage despite variations in the bias voltage on the base lead.

emphasis   The passage of a signal through a network that emphasizes one range of frequencies over others. Also called *preemphasis.*

energized   (1) Electrically connected to a voltage source. (2) Hot. (3) Live. (4) Alive.

energy   Abbreviated *E* or *W*, the capacity for doing work. Energy may be transferred from one form to another (chemical, thermal, nuclear, electrical), but it cannot be created or destroyed.

envelope   (1) The glass or metal housing of an electron tube. Also called a *bulb.* (2) A curve drawn to pass through the positive and negative peaks of a modulated RF carrier signal.

epsilon (ε)   The fifth letter of the Greek alphabet, used to represent the symbol for *dielectric constant* or *permitivity* (capacity).

equalizer   A passive or active network that compensates for any desired amplitude- or phase-frequency response of a system or component.

equal ripple filter   See *Chebyshev filter.*

erg   The work done by a force of 1 dyne applied through a distance of 1 centimeter. Equal to 10^{-7} joules.

ERP   Abbreviation for *effective radiated power.*

Esaki diode   See *tunnel diode.*

eta (η)   The seventh letter of the Greek alphabet, used to represent *efficiency,* and the *intrinsic standoff ratio* of a unijunction transistor.

eV   Abbreviation for *electronvolt.*

E-vector   A vector that represents the electric field of an electromagnetic wave, which is perpendicular to the H-vector and the direction of propagation in free space.

E-wave   British term for *transverse magnetic (TM) wave.*

exa   Abbreviated *E,* prefix for 10^{18}.

excitation   (1) The signal voltage that is applied to the control terminal, such as the base, gate, or grid of an active amplifying device. Also called *drive.* (2) The application of signal power to a transmitting antenna.

exclusive-OR gate   Abbreviated *XOR gate,* a logic gate whose output is at logic 1 when only one, but not more than one of its inputs is at logic 1; otherwise its output is at logic 0, signifying that all inputs are at the same logic level. Also called a *comparison gate.*

Exclusive-OR gate

exp   The abbreviation for the natural or Napierian base *e* (2.71828). As written in equations, the quantity immediately following it is to be treated as the exponent of *e,* so that $\exp(-0.6) = e^{-0.6}$.

expander Also spelled expandor, the part of a compander that is used as the receiving end of a circuit to return the compressed signal to its original form. It attenuates weak signals and amplifies strong signals. Its function is the opposite of a compressor.

extended-cutoff tube See *remote-cutoff tube*.

extremely high frequency Abbreviated *EHF*, any frequency in the region from 30 to 300 GHz.

extremely low frequency Abbreviated *ELF*, any frequency below 300 Hz.

f (1) Prefix symbol for *femto*, 10^{-15}. (2) Mathematical symbol for *frequency*.

F (1) Symbol for *filament, Faraday constant*. (2) Unit symbol for *farad, Fahrenheit*. (3) Letter designation for *fuse* on schematic diagrams.

facsimile Abbreviated *FAX*, the process whereby fixed graphic material is sent or received by electrical signals to produce a permanent record of the likeness (facsimile) of the source copy.

fade To gradually change the amplitude of a signal or the response of a circuit.

fade-in To gradually increase the amplitude of a signal or the response of a circuit.

fade-out To gradually decrease the amplitude of a signal or the response of a circuit.

Fahrenheit temperature scale Symbolized by *F*, a temperature scale in which the freezing point of water is 32 degrees while the boiling point of water is 212 degrees at the atmospheric pressure of 760 millimeters of mercury. In terms of a centigrade (Celsius) temperature, the equivalent Fahrenheit temperature is computed from °F = (9/5)°C + 32.

fall time The amount of time for a waveform to decrease from 90 to 10 percent of its maximum amplitude.

farad Abbreviated *F*, the unit of capacitance equal to a charge of 1 coulomb on a capacitor with a potential difference of 1 volt across it.

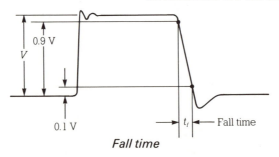

Fall time

faraday A unit of charge equal to 96 484 coulombs.

Faraday constant Symbolized by the letter *F*, equal to 9.649×10^4 coulombs per mole.

Faraday, Michael An English chemist and physicist (1791–1867) who developed the dynamo, discovered electromagnetic induction, and after whom the units of electrical charge (faraday) and capacitance (farad) were named.

Faraday's law of induction The induced electromotive force in a circuit is equal to the time rate of change of the magnetic flux through the circuit, or:

$$ \text{EMF} = -\frac{\partial \phi}{\partial t} $$

Also known as the *law of electromagnetic induction.*

Faraday shield An electrostatic shield formed by a wire screen connected to a common conductor grounded at one end. Also called a Faraday screen or Faraday-cage.

far field Any electromagnetic field generally greater than a distance of one wavelength from its source.

fault current A current that flows as a result of a failure in an electrical system or component.

FAX (or fax) Abbreviation for *facsimile.*

fc Abbreviation for *foot-candle.*

FCC Abbreviation for Federal Communications Commission, a U.S. government agency that regulates radio communications.

feedback The return of a portion or all of the output of a circuit or device to its input. Positive feedback requires that the signal be fed back in phase with the input, which in turn increases circuit amplification and speed but may cause oscillation. With negative feedback, the signal fed back is 180 degrees out of phase with the input, which in turn decreases amplification but stabilizes circuit performance.

feedback control system A control system having one or more feedback control loops.

feedback path In a feedback control loop, the transmission path from the output to the input of the loop.

feeder In a power or transmission system, a conductor or group of conductors connecting two different generation or distribution units.

feedthrough A conductor used to pass signals through one side of a panel to the other.

feedthrough capacitor A feedthrough insulator which provides a specific capacitance value between the feedthrough conductor and a metal chassis or panel through which the conductor is passing.

feedthrough insulator An insulator which allows a wire or cable to be fed through walls with a minimum of leakage current.

female connector A recessed connector, usually a jack or socket, which mates with a corresponding male connector.

femto Abbreviated *f*, prefix for 10^{-15}.

ferrite A type of ferric oxide material having magnetic properties with high resistivity that is used in transformer cores, resulting in low eddy current losses.

51

ferroelectric Pertaining to a characteristic of certain metallic materials, in which they are polarized in a given direction by the application of a polarized electric field of sufficient intensity. Once polarized, the material retains its electrical polarization even after the electric field is removed.

ferromagnetic Pertaining to a characteristic of certain metallic materials, in which they are polarized in a given direction by the application of a polarized magnetic field of sufficient intensity. Once polarized, the material retains its magnetic polarization even after the magnetic field is removed.

FET Acronym for *field effect transistor.*

FF Abbreviation for *flip-flop.*

fiber optics The technology of transmitting light through long, thin, flexible fibers of glass, plastic, or other transparent materials.

field (1) A general term which refers to the region under the influence of some physical phenomenon, such as electricity or magnetism. (2) The field coil of a motor or generator.

field coil A coil of insulated wire wound on an iron core which produces a magnetic field when current flows in the coil windings.

field effect transistor Abbreviated *FET.* See *junction field effect transistor.*

field intensity See *field strength.*

field pattern See *radiation pattern.*

field strength (1) The effective value of the electric field intensity expressed in volts per meter at a given distance from the source. Also called *field intensity.* (2) The magnetic flux developed at a given point by either an electromagnetic or permanent magnet.

FIFO Acronym for *first-in first-out.*

figure of merit A quantitative characteristic, such as gain, referring to the suitability of a device for a particular application.

filament (1) Symbolized by the letter *F* on schematic diagrams, a cathode made of resistance wire or ribbon through which an electric current is passed to produce the high temperatures required for electron emission in a thermionic tube. Also called a *heater.* (2) A directly heated cathode.

filter A selective electrical network, either active or passive, which allows the passage of a band of frequencies while rejecting the passage of another band of frequencies.

filter capacitor A capacitor used in a filter circuit, generally those of power supplies.

filter choke An iron-core inductor which allows direct current to pass but stops pulsating alternating current.

filter passband See *passband.*

filter section Any simple network which may be connected in cascade to form a filter.

fin A metal disc or thin metal strip mechanically attached to a solid-state device for the purpose of dissipating heat.

final The power amplifier stage of an audio amplifier or transmitter which feeds the load. Also called the final amplifier.

firing potential The voltage at which conduction through a gas-filled tube begins.

first detector (1) The stage of a superheterodyne receiver where a local oscillator signal is combined with the incoming modulated radio frequency signal to produce a modulated intermediate frequency signal. (2) A mixer.

first-in first-out Abbreviated *FIFO*, a method which establishes the sequence in which digital data is taken from a shift register or memory element.

fishpaper A tough but flexible insulating material used to separate transformer coil windings from the core, the conductor from the armature, or field coils from the poles.

fixed capacitor A capacitor that has a definite value of capacitance which cannot be changed.

fixed resistor A resistor that has a definite value of resistance which cannot be changed.

FL Abbreviation for *full load*.

flashover An electric discharge around or over the surface of an insulator.

Fleming's rules Two rules, also individually called the *right-hand rule* and the *left-hand rule*. The right-hand rule states that if the thumb, index, and middle fingers of the right hand are extended at right angles to each other with the index finger representing the direction of magnetic lines of force and the middle finger representing the direction of current flow, then the thumb will be pointing in the direction of motion of a movable current-carrying wire or an electron beam in a magnetic field. The left-

hand rule states that if the thumb, index, and middle fingers of the left hand are extended at right angles to each other with the index finger representing the direction of magnetic lines of force and the middle finger representing the direction of electron flow, then the thumb will be pointing in the direction of motion of a movable current-carrying wire or an electron beam in a magnetic field. *(See figure on page 54.)*

flicker (1) In a cathode-ray tube, the uncontrolled movement in the trace pattern when the field frequency is not sufficient for synchronization of the pattern. (2) Noise in an amplifier having a frequency higher than drift but lower than that of the power line frequency. Also called *jitter* or *wobble*.

flip-flop Abbreviated *FF*, a bistable multivibrator with a trigger input and an output that has two stable states. The input trigger changes the circuit from one state to the other.

floating The condition whereby a circuit or input lead is not grounded and not tied to a source of potential.

floating input An input of an amplifier or other circuit which is not connected to the circuit ground.

floating neutral A circuit in which the voltage to ground varies with circuit conditions.

float switch A switch actuated by a float at the surface of a liquid.

flourescent lamp An electric discharge lamp in which an ionizing gas activates a flourescent material inside a glass envelope, which in turn produces radiation.

flow The movement of electric charge, gas, or liquid.

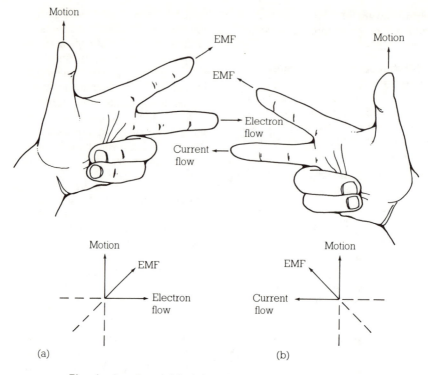

Motion

EMF

EMF

Electron flow

Current flow

Motion

Motion

EMF

Electron flow

EMF

Current flow

Motion

(a)

(b)

Fleming's rules: (a) Left-hand rule; (b) Right-hand rule

flutter (1) A type of distortion that occurs in sound reproduction systems as a result of undesired sound variations during the recording, duplication, or reproduction process. Also called *wow*. (2) A fast-changing variation in received signal strength caused by antenna movements or interaction with a signal on another frequency.

flux (1) Symbolized by the Greek letter ϕ, the number of lines of force that pass through a given surface in a field. (2) The flow of a quantity, such as mass or energy, that passes through a surface per unit time. (3) A material used to remove oxide films from metallic surfaces in preparation for brazing, soldering, or welding.

flux density Symbolized by the letter D, the quantity of magnetic lines of force (flux) passing through a unit area. Also called *induction density*.

flux linkage Symbolized by the Greek letter ψ, the product of the number of turns of a coil of wire and the magnetic flux through it, so that $\psi = N\phi$.

flyback The return of the electron beam in a cathode-ray tube to its starting point after a sweep across the screen. Also called *retrace*.

flyback time The time period during which an electron beam returns from the end of a scanning line to the beginning of the next line.

flyback transformer A transformer used in the horizontal deflection circuit of a television receiver as well as providing the filament voltage for the high-voltage rectifier. Also called a *horizontal output transformer.*

flywheel effect The ability of a resonant circuit to maintain oscillation at a constant frequency when fed with short pulses of energy at constant frequency and phase.

FM (or fm) Abbreviation for *frequency modulation.*

FM band For FM broadcasting, the frequency range from 88.1 to 107.9 MHz. It is divided into 200 channels, each having a bandwidth of 200 kHz.

focusing coil The coil around the neck of a cathode-ray tube. It provides a magnetic field parallel to the electron beam and controls the cross-sectional area of the beam on the screen.

foldback See *current foldback.*

folded dipole A dipole antenna whose ends are folded back and joined together at the center. The impedance is about 300 ohms, compared with about 70 ohms for a simple dipole.

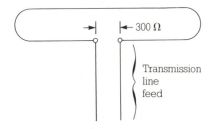

300 Ω

Transmission line feed

Folded dipole

follower (1) An amplifier circuit whose output, without phase reversal, is essentially the same in intensity as its input. (2) A buffer.

foot-candle Abbreviated *fc*, a measure of illumination equal to that produced on 1 square foot of a surface at a distance of 1 foot from a standard candle. Equal to 10.764 lumens per square meter.

forced oscillation The resulting oscillation when a system that is capable of free oscillation is subjected to an oscillatory driving force.

form To apply a voltage to a component, such as an electrolytic capacitor, in the manufacturing process to cause a permanent change in its characteristics.

form factor The ratio of the root mean square, or effective value, of an AC signal to the average value during a half-cycle. For an inductor, a factor that depends on the ratio of the diameter to the length of the coil.

forward biased A bias voltage applied to a semiconductor junction in the direction that causes a large current flow.

forward direction The direction of least resistance to current flow in a diode or semiconductor device.

forward path In a feedback control loop, the transmission path from the input to output.

forward power See *incident power.*

forward resistance The resistance measured at a specified forward voltage drop and current in a diode or semiconductor device.

forward voltage A voltage having the correct polarity to send current through a semiconductor device in the forward direction.

Foucault current See *eddy current.*

Fourier, Baron Jean Baptiste Joseph   A French mathematician and physicist (1768–1830) who was noted for his work on heat and numerical equations.

Fourier series   An infinite series expressing a periodic waveform as a DC term plus sine and cosine waveforms whose frequencies are harmonic multiples of the periodic waveform's fundamental frequency, so that

$$f(t) = \frac{1}{2}A_o + \sum_{n=1}^{\infty}\left[A_n\cos(n\omega t) \right.$$

$$\left. + B_n\sin(n\omega t) \right]$$

where:

$$A_n = \frac{1}{\pi}\int_0^{2\pi} f(t)\cos(n\omega t)d(\omega t)$$

$$B_n = \frac{1}{\pi}\int_0^{2\pi} f(t)\sin(n\omega t)d(\omega t)$$

Fourier transform   An analytical method for representing a transient or aperiodic signal as a continuous frequency spectrum, so that

$$F(\omega) = \int_{-\infty}^{+\infty} f(t)e^{-j\omega t}\, dt$$

and the inverse Fourier transform is given by:

$$f(t) = \frac{1}{2\pi}\int_{-\infty}^{+\infty} F(\omega)e^{j\omega t}\, dt$$

Frahm frequency meter   See *vibrating-reed frequency meter.*

free impedance   The input impedance of a transducer with no output load.

free-running frequency   The frequency at which a normally synchronized oscillator operates in the absence of a synchronizing signal.

free-wheel rectifier   A diode connected in reverse polarity across an inductive load, such as a DC relay, in order to provide a path for the resulting transient current when the source current is removed.

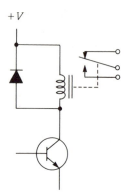

Free-wheel rectifier

freq   Abbreviation for *frequency.*

frequency   Symbolized by *f* and the Greek letter ν, and abbreviated *freq,* the number of cycles of a periodic event per unit of time. Electrical frequency is expressed in cycles per second, but is now given the unit of hertz. Angular frequency, symbolized by the Greek letter ω, is expressed in radians per second.

frequency band   A specific and continuous range of frequencies.

frequency converter   (1) A device whose input and output frequencies are different, such as a full-wave rectifier or a mixer. (2) A frequency changer.

frequency compensation   In an amplifier, a method whereby the passband gain of an amplifier is extended higher, lower, or both. Also called *compensation.*

frequency cutoff   See *cutoff frequency.*

frequency departure The difference between the actual carrier frequency and its desired value. Also called *frequency drift.*

frequency deviation In a frequency modulation system, the difference between the instantaneous frequency of the modulated signal and its carrier frequency. Also called *deviation.*

frequency discriminator A circuit that converts a frequency-modulated signal into an audio signal. An FM demodulator.

frequency distortion The type of distortion that results when all frequencies of a complex waveform are not amplified or attenuated equally. Also called *amplitude distortion.*

frequency drift See *frequency departure.*

frequency frogging An exchange of carrier frequencies between the input and output of each of a system of repeaters. This results in greater system stability, less crosstalk, and virtually no need for equalization. For each repeater, low-frequency carriers are translated upward while high-frequency carriers are translated downward.

frequency modulation Abbreviated *FM,* modulation of a sinewave carrier so that the carrier frequency is proportional to the amplitude of the modulating signal.

frequency modulation deviation The maximum difference between the instantaneous frequency of a modulated signal and the carrier frequency.

frequency multiplex The transmission of two or more signals with different frequencies over a common wire or path.

frequency multiplier A circuit whose output frequency is an integral multiple, or harmonic, of its input frequency.

frequency offset The difference between a steady-state frequency and a given reference frequency.

frequency pulling The change in an oscillator's output frequency as a result of changing load impedance.

frequency response A measure of how effectively a circuit passes a signal as a function of frequency.

frequency-shift keying Abbreviated *FSK,* a form of frequency modulation in which the carrier frequency shifts between two predetermined values.

frequency stability The ability of electronic equipment to maintain its desired operating frequency over wide external changes, such as in its load and ambient temperature.

frequency standard A highly stable and accurate source of known frequency used for calibration or comparison.

frequency synthesizer A device, generally built around a phase-locked loop, that provides a switch-selectable choice of a large number of output frequencies derived from one or more crystal-controlled oscillators.

frequency-to-voltage converter Abbreviated *F/V converter,* a circuit that provides a DC voltage that is proportional to its input frequency.

Fresnel A unit of frequency equal to 1 terahertz, 10^{12} Hz.

fringe area An area just beyond the limits of the reliable reception area of a television or radio antenna where the signals are weak and erratic.

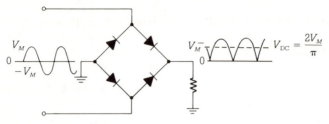

Full-wave rectifier (bridge type)

fringe effect   The extension of the electrostatic field of an air capacitor beyond the space between its parallel plates.

fringe howl   A howl or squeal noise heard when a circuit in a receiver is on the verge of oscillation.

front end   (1) The first stage of a radio receiver. Sometimes called a *radio-frequency (RF) amplifier.* (2) The high-gain, low-noise stage of a tuner or receiver used to select the desired station and convert the radio frequency signal to an intermediate frequency.

front-end rejection   The ability, expressed in decibels, of a receiver to reject those signals outside its tuned bandwidth.

front-to-back ratio   The ratio of an antenna's power gain or signal strength transmitted in the forward, or front, direction to that transmitted in the reverse, or back, direction.

FS   Abbreviation for *full scale.*

FSK (or fsk)   Abbreviation for *frequency-shift keying.*

full load   Abbreviated *FL*, the greatest load that a circuit or device is designed to handle under specified conditions.

full scale   Abbreviated *FS*, the maximum possible scale reading of a meter or digital display.

full-wave   (1) Having an electrical length of one wavelength. (2) Pertaining to the entire cycle of an AC waveform.

full-wave rectifier   A diode circuit where both the positive and negative cycles of an AC waveform are allowed to pass through. In this manner, an AC voltage is converted into a pulsed DC voltage.

function generator   (1) A device whose output can be changed in terms of frequency, amplitude, and waveshape. (2) A waveform generator.

fundamental frequency   (1) The lowest frequency component of an AC signal. (2) The lowest frequency at which a system can freely oscillate.

fuse   Symbolized by the letter F on schematic diagrams, usually a short piece of wire or metal which melts and thus breaks an electric circuit when the current through it exceeds a rated value.

fuse block   An insulated base on which a fuse holder or other contacts for holding fuses are mounted.

F/V converter   Abbreviation for *frequency-to-voltage converter.*

G (1) Mathematical symbol for *conductance*. (2) Abbreviation for *gate, grid,* or *ground.* (3) Letter designation for a *generator* on schematic diagrams. (4) Prefix symbol for *giga,* 10^9. (5) Unit symbol for *gauss.*

gage (1) Also spelled gauge, an instrument for measurement or testing. (2) A transducer.

gage factor Symbolized by the letter K, a dimensionless quantity that is the ratio of the relative change in resistance of a strain gage transducer to the relative change in the physical quantity being measured, such as length or strain, and is computed from

$$K = \frac{\Delta R/R}{\Delta L/L}$$

gain Symbolized by the letter A, an increase in signal intensity, expressed as a ratio of output to input, or between any two points in a circuit. See also *current gain, power gain,* and *voltage gain.*

gain-bandwidth product Abbreviated *GBP,* the product of the closed-loop gain of an operational amplifier and its corresponding bandwidth, which is a constant value.

gain-margin As a measure of operational amplifier feedback stability, the amount of gain needed to cause oscillation.

galvanometer An instrument for measuring the polarity and magnitude of small electric currents.

gamma (Γ, or γ) Third letter of the Greek alphabet, used to represent *permeance* (γ), or *conductivity* (Γ).

gamma rays Electromagnetic radiation similar to X-rays, but having shorter wavelengths, approximately 5×10^{-9} to 5×10^{-11} cm.

gang To mechanically couple or connect a set of similar components, such as potentiometers, capacitors, or switches, so that they can all be operated simultaneously.

gate Abbreviated *G,* the control terminal of a field effect transistor, thyristor, or silicon controlled rectifier.

gauss Abbreviated G, a unit of magnetic induction (magnetic field intensity) in the CGS system. Equal to 1 maxwell per square centimeter, 10^{-4} webers per square meter.

Gauss, Karl Friedrich A German mathematician and astronomer (1777–1855) who pioneered in applying mathematical theory to electricity and magnetism, and after whom the unit of magnetic field intensity was named (now known as the oersted). Since 1932 the gauss has been used as a unit of magnetic induction.

Gaussian noise Random electrical noise having a statistical distribution the same as a normal distribution curve.

gaussmeter An instrument that measures magnetic field, or flux, density by its interaction with an internal magnetic field.

Gauss's law The surface integral of the normal component of the electric flux density D over any closed surface equals the charge enclosed, so that for a sphere of radius r, $Q = 4\pi\epsilon Dr^2$.

Gb Unity symbol for *gilbert*.

GBP Abbreviation for *gain-bandwidth product*.

GDO Abbreviation for *grid-dip meter*.

Ge Chemical symbol for *germanium*.

genemotor A dynamotor having two armature windings. One winding serves as the driving motor and generally operates from a battery. The other winding functions as a high-voltage DC generator.

general-purpose motor Any motor generally taken to be less than 200 horsepower but with speeds greater than 450 revolutions per minute and rated for continuous operation with no restriction as to its application.

generated noise The noise due to such causes as the electromotive force generated by the rubbing together of two dissimilar metals, or the electromotive force generated from thermoelectric effects at points where dissimilar metals are joined.

generator (1) Symbolized by the letter G on schematic diagrams, a rotating machine which converts mechanical energy into electrical energy. (2) An electronic device which produces waveforms of specific shape, amplitude, and frequency.

geometric mean The Nth root of the product of a series of N quantities, so that

$$\overline{X} = \sqrt[N]{X_1 X_2 X_3 \ldots X_N}$$

germanium Symbolized by Ge, a gray-white, brittle, metallic element having semiconductor properties that is used in the manufacture of transistors and diodes.

GFI Abbreviation for *ground fault interrupter*.

GHz Unit symbol for gigahertz.

giga Abbreviated G, prefix for 10^9.

gilbert Abbreviated Gb, a unit of magnetomotive force in the CGS system. It is equal to the force required to produce one maxwell of magnetic flux in a magnetic circuit of unit reluctance. Equal to 0.7958 ampere-turn, or 1 oersted-centimeter.

gimmick A type of capacitor formed by twisting two insulated wires together, or by bringing two parallel conductors into close proximity.

glitch A slang term used to refer to any form of interference or noise that interferes with the operation of a circuit or system.

g_m Symbol used to designate the *mutual conductance,* or *transconductance* of either a vacuum tube or field effect transistor.

GND (or gnd) Abbreviation for *ground.*

governor An attachment to a motor that automatically limits or controls the speed of the motor.

grad Abbreviation for *gradient.*

gradient (1) Abbreviated *grad,* a rate of change of a quantity with respect to distance or time. (2) A vector operator ∇ in which

$$\nabla = \mathbf{i}\frac{\partial}{\partial x} + \mathbf{j}\frac{\partial}{\partial y} + \mathbf{k}\frac{\partial}{\partial z}$$

so that, for a scalar quantity U, the gradient, or grad U is

$$\nabla U = \mathbf{i}\frac{\partial U}{\partial x} + \mathbf{j}\frac{\partial U}{\partial y} + \mathbf{k}\frac{\partial U}{\partial z}$$

grass Random noise pulses that produce a series of small vertical fluctuations near the baseline of an oscilloscope or radar display. Also called *hash.*

graticule A calibrated screen placed in front of an oscilloscope display for measuring the amount of deflection of the beam.

grid (1) Abbreviated *G,* the control element of a vacuum tube in the form of either a wire or screen placed between the plate (i.e., anode) and cathode. (2) A two-dimensional network of equally spaced parallel lines superimposed at right angles on another set of equally spaced parallel lines, thereby forming square areas. (3) An interconnected system in which high-voltage, high-current lines overlay and are connected with networks of lower voltages, such as a power grid.

grid-dip meter Abbreviated *GDO,* a vacuum tube oscillator with a meter that reads the average value of the oscillator's grid current. It is used primarily as an absorption wavemeter so that when it is coupled to a resonant system, and its own oscillator frequency is turned to it, energy is then drawn from the oscillator and it becomes loaded down so that the grid current decreases or dips. Also called a grid-dip oscillator.

grid-leak A high resistance connected between the grid and cathode of a vacuum tube so that the charges that accumulate on the grid are allowed to be diverted, or leak off, to ground.

ground (1) Abbreviated *G* or *GND,* a point considered to be at zero potential and to which all other potentials are referred. (2) Earth. (3) To connect a point or circuit lead to ground. (4) To short a signal path to ground.

grounded-base amplifier See *common-base amplifier.*

grounded-emitter amplifier See *common-emitter amplifier.*

grounded-gate amplifier See *common-gate amplifier.*

grounded-source amplifier See *common-source amplifier.*

ground fault interrupter Abbreviated *GFI*, a circuit breaker which is activated when a ground fault current exceeds a relatively small value, such as 10 microamperes.

ground loop The generation of noise or undesirable signals within a ground path or conductor as a result of current within this path originating from another voltage source. This frequently results from connecting two separate grounds to a circuit.

ground plane (1) An omnidirectional, low radiation angle antenna that is vertically polarized and uses the earth's (ground) surface (plane) as part of its radiating system. (2) A metallic sheet, usually copper, used in testing so that the actual grounding conditions may be approximated.

ground return (1) In a circuit, a conductor between two points whose path is completed through ground. (2) In radar systems, the echos

reflected from the earth's surface as well as from fixed objects on it.

ground rod A pipe, usually copper or steel, driven into the ground to make an electrical connection with it.

ground wave A radio wave that travels from the transmitting antenna along the earth's surface rather than through the atmosphere.

ground wire A conductor leading to an electrical/mechanical connection with the earth.

guard ring A metal ring electrode used to limit the extent of an electric field, or to distribute the charge uniformly over the surface of an object.

guard shield A metallic shield surrounding the input circuit or stage of an amplifier.

gutta-percha A natural vegetable gum similar to rubber that is used primarily as insulation for wires and cables.

gyrator A directional phase changer in which the changes in phase in opposite directions differ by 180 degrees.

h (1) Mathematical symbol for *Planck's constant.* (2) Prefix symbol for *hecto,* 10^2.

H (1) Mathematical symbol for *magnetic field strength.* (2) Abbreviation for *heater.* (3) Unity symbol for *henry.*

half-cycle The interval for a periodic waveform to complete one-half of its cycle, or 180 degrees.

half-power frequency (1) The frequency at which the power gain of a system decreases to one-half of its maximum value. Also called *half-power point.* (2) The frequency at which the amplitude response of a system is 3 dB less than its maximum value.

half-power points See *half-power frequency.*

half-wave (1) Having an electrical length of one-half wavelength. (2) Pertaining to one-half of a cycle of a waveform.

half-wave dipole A dipole antenna whose electrical length at the frequency of interest is one-half wavelength.

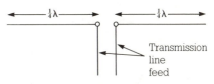

Half-wave dipole

half-wave rectifier A diode circuit where only one-half cycle of the input waveform, either positive or negative, is allowed to pass through. In this manner, an AC voltage is converted into a pulsed DC voltage. *(See figure on page 64.)*

Hall effect When a current is passed through a semiconductor bar situated within a magnetic field that is perpendicular to the direction of the current, a voltage is developed across the bar that is perpendicular to both the magnetic field and current. *(See figure on page 64.)*

harmonic A periodic waveform whose frequency is an integral multiple of some fundamental frequency.

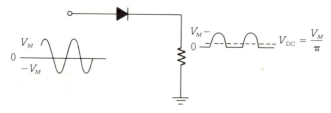

Half-wave rectifier

harmonic distortion   A form of distortion caused by a signal passing through a nonlinear system whereby harmonics are added to the fundamental signal.

harmonic mean   The reciprocal of the arithmetic mean of the reciprocals of the numbers, so that

$$\overline{X} = \cfrac{1}{\cfrac{1}{N}\sum \cfrac{1}{X_i}}$$

$$= \cfrac{N}{\cfrac{1}{X_1} + \cfrac{1}{X_2} + \cfrac{1}{X_3} \cdots + \cfrac{1}{X_N}}$$

harmonic oscillator   An oscillator whose output frequency and amplitude are essentially constant, but whose output closely approximates a sinewave so that harmonic multiples are produced.

Hartley oscillator   A sinewave oscillator, built around either a tube or transistor, whose feedback loop circuit contains a tapped inductor in parallel with a single capacitor. The inductor and capacitor values determine the oscillator's frequency.

hash   Electrical noise that results from contacts that open and close rapidly. Also called *grass*.

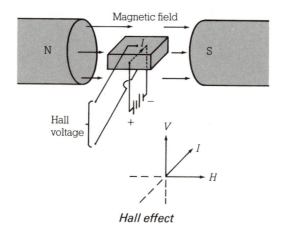

Hall effect

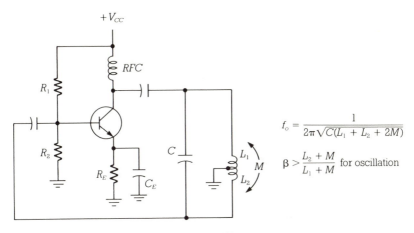

$$f_o = \frac{1}{2\pi\sqrt{C(L_1 + L_2 + 2M)}}$$

$$\beta > \frac{L_2 + M}{L_1 + M} \text{ for oscillation}$$

Hartley oscillator

Hay bridge An AC bridge for measuring inductance in terms of resistance, capacitance, and frequency; best suited for high-Q inductors.

heater Abbreviated *H*. (1) The filament in a vacuum tube. Also called a *filament*. (2) An element that supplies the heat to an indirectly heated cathode. (3) A wire-type resistor that converts electrical energy into heat.

heat sink A metallic mass that conducts, convects, radiates, or otherwise dissipates heat from a semiconductor device to the surrounding air.

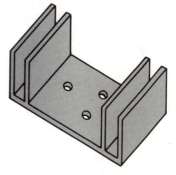

Heat sink

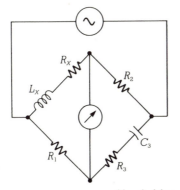

$$L_X = \frac{R_1 R_2 C_3}{1 + (\omega C_3 R_3)^2}$$

$$Q_X = \frac{\omega L_X}{R_X} = \frac{1}{\omega C_3 R_3}$$

Hay bridge

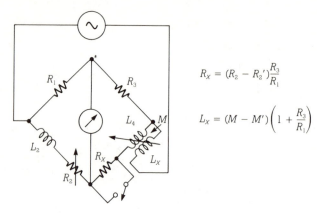

$$R_X = (R_2 - R_2')\frac{R_3}{R_1}$$

$$L_X = (M - M')\left(1 + \frac{R_3}{R_1}\right)$$

Heaviside-Campbell mutual inductance bridge

Heaviside-Campbell mutual inductance bridge An AC bridge network, a variation of the Heaviside mutual inductance bridge, used for the comparison of mutual inductances.

Heaviside mutual inductance bridge An AC bridge network whose balance is independent of frequency, used for the comparison of self- and mutual inductances.

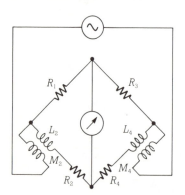

Heaviside inductance bridge

hecto Abbreviated h, prefix for 10^2.

henry Abbreviated H or Hy, the unit for inductance equal to when a coun-

terelectromotive force of 1 volt is generated when a current is changing at the rate of 1 ampere per second.

Henry, Joseph An American physicist (1798–1878) who discovered self-inductance, improved the electromagnet, and after whom the unit of inductance was named. Independently of Faraday, Henry discovered the principle of induced current.

heptode A class of vacuum tubes that contains seven elements: an anode, a cathode, and five grids.

hertz Abbreviated Hz, the unit of frequency, equal to 1 cycle per second.

Hertz antenna A dipole antenna.

Hertz, Heinrich Rudolf A German physicist (1857–1894) who confirmed Maxwell's electromagnetic theory, demonstrated that electromagnetic (radio) waves travel at the speed of light and can be reflected, refracted, and polarized like light, and after whom the unit of frequency was named.

Hertzian wave   See *radio wave.*

heterodyne   (1) To mix two AC signals of different frequencies in a nonlinear device. The spectrum of the output waveform then consists of components of the two input frequencies plus the sum and difference frequencies of the two inputs. Also called *mix.* (2) To beat.

heterodyne frequency   The intermediate (sum or difference) frequency produced by mixing two other frequencies.

heterodyne oscillator   An oscillator that produces an output frequency by mixing two other frequencies.

heterodyne wavemeter   A wavemeter using the heterodyne (mixing) principle to compare the measured frequency with one produced by a calibrated oscillator.

high band   Generally the frequency range occupied by television channels 7 through 13, or 174–216 MHz.

high-fidelity   A system capable of reproducing audio signals with a minimum of distortion. Also called *Hi-Fi.*

high frequency   Abbreviated *HF,* any frequency in the region from 3 to 30 MHz.

high-pass filter   (1) A circuit which passes all frequencies without attenuation above a specified frequency while rejecting all frequencies below the specified frequency. (2) A lead network.

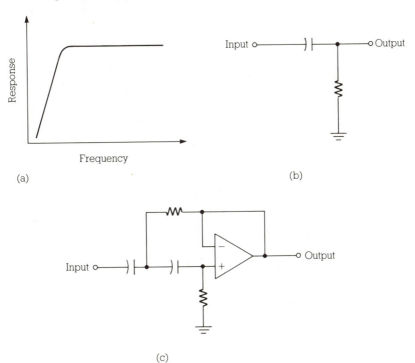

High-pass filter: (a) Frequency response; (b) Passive circuit; (c) Active circuit

high tension High voltage, generally at or above levels which cause death.

high voltage Abbreviated *HV*, a relative term used to refer to an abnormally high voltage. In AC power distribution systems, it is generally used for voltages greater than 13 kV.

Hi-Fi (or hi-fi) Acronym for *high-fidelity*.

H-network A two-port, five-branch network arranged in the form of the letter H lying on its side.

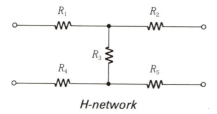

H-network

holding current (1) The minimum current in a relay coil that must be maintained to keep the relay energized. (2) In a silicon controlled rectifier or neon tube, the minimum current required to maintain conduction.

hookup (1) The method of connection of the various units of a circuit, i.e., a schematic diagram. (2) The circuit itself.

hookup wire Soft-drawn, copper wire, solid or stranded from 18 to 20 gauge, used to connect circuits together.

horizontal amplifier An amplifier in an oscilloscope for signals intended to produce a horizontal deflection. Also called an *x-axis amplifier*.

horizontal output transformer See *flyback transformer.*

horsepower Abbreviated *hp*, a unit of power equal to 33 000 foot-pounds per minute, or 746 watts.

hot (1) Connected to a source of voltage or current. (2) Energized. (3) Not grounded. (3) At a high temperature.

hot-carrier diode See *Schottky diode.*

hp Abbreviation for *horsepower.*

h-parameters Abbreviation for *hybrid parameters.*

hum A low-pitch audible noise resulting from an AC power supply, the ripple voltage from a DC power supply, or by induction from nearby power systems. Generally hum has a frequency equal to the power line or its second harmonic.

humming A low frequency sound, generally 120 Hz, produced by a transformer having either loose laminations or magnetization effects in its iron core.

HV Abbreviation for *high voltage.*

H-vector A vector quantity, representing the magnetic field of an electromagnetic wave, which in free space is perpendicular to the electric field and the direction of propagation.

H-wave British term for *transverse electric (TE) wave.*

Hy (or hy) Abbreviation for *henry.*

hybrid circuit Any circuit having two different components that perform the same function, such as a circuit containing both vacuum tubes and transistors.

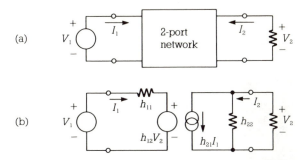

Hybrid-parameters: (a) Two-port network; (b) Equivalent circuit

hybrid parameters Abbreviated *h-parameters,* the four parameters of a two-port network, or transistor, relating input and output voltages and currents so that

$$h_{11} = \frac{V_1}{I_1} = \text{input impedance}$$

when $V_2 = 0$

$$h_{12} = \frac{V_1}{V_2} = \text{reverse voltage transfer}$$

ratio when $I_1 = 0$

$$h_{21} = \frac{I_2}{I_1} = \text{forward current transfer}$$

ratio when $V_2 = 0$

$$h_{22} = \frac{I_2}{V_2} = \text{output admittance}$$

when $I_1 = 0$

hysteresis The property of a system, which may be electromagnetic, electronic (e.g., Schmitt trigger), or mechanical, shown by the nonlinear dependence of the value of the output, for a given excursion of the input, upon the history of prior behavior of the system.

hysteresis curve A nonlinear, double-valued curve of the relationship between two variables. In an electromagnetic circuit, a hysteresis curve shows the relation between the magnetic force H and the resultant magnetic flux B. When the magnetic force is allowed to take on positive as well as negative polarities, the resulting closed loop is called a hysteresis loop.

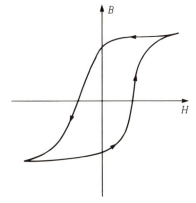

Hysteresis curve

Hz Abbreviation for *hertz.*

i (1) Mathematical symbol for *AC current*. (2) A complex operator, used by other than electrical engineers, equal to the square root of − 1.

I Mathematical symbol for *DC current*.

IA Abbreviation for *instrumentation amplifier*.

IC Abbreviation for *integrated circuit*.

I channel A 1.5-MHz wide channel used in the American NSTC color television system for transmitting cyanorange color information. The signals in this channel are known as I signals.

idler frequency The frequency at which a parametric amplifier produces its amplified output.

IF Abbreviation for *intermediate frequency*.

IGFET Acronym for *insulated gate field effect transistor*.

IM Abbreviation for *intermodulation*.

image frequency An unwanted carrier frequency in a superheterodyne receiver that differs from the tuned frequency by twice the intermediate frequency.

image frequency rejection ratio In a superheterodyne receiver, the ratio of the receiver response at the desired frequency to its response at the image frequency.

image impedance The load impedance that will permit maximum power transfer when connected across a device. For a device having a complex output impedance, the image impedance is equal to the complex conjugate of the device's output impedance.

imaginary number A number whose square is a negative number.

immitance An acronym for *impedance* and *admittance*, a term used to denote both impedance and admittance.

impedance Symbolized by the letter Z, the total opposition in ohms that a circuit offers to the flow of alternating current. It consists of the combined effects of resistance, inductive, and capacitive reactance, so that $Z = R + j(X_L - X_C)$. In terms of AC voltage and current, impedance is the ratio of voltage to current.

impedance angle See *phase angle*.

impedance bridge A bridge network used to measure impedance.

impedance matching The process whereby the load impedance is adjusted to the output impedance of a network for maximum power transfer from its source to the load.

impedance triangle A right-triangle diagram with sides proportional to the resistance and reactance of an AC circuit. The triangle's hypotenuse then represents the impedance of the circuit. The cosine of the angle between the hypotenuse and resistance lines equals the power factor of the circuit, while the arctangent of the ratio of the reactance to resistance lines equals the phase angle.

impulse A pulse so short that it may be considered as having an infinitesimally small duration.

impulse noise Noise characterized by transient, short-duration disturbances that are distributed uniformly over the passband of the transmission system.

incidence angle The angle between an approaching beam of radiation and a line perpendicular to the surface in the path of the beam.

incident power In a transmission line, the product of the magnitude of the voltage and wave components traveling in the direction from the source to the antenna or load. Also called *forward power*.

incident wave (1) A wave that impinges on a discontinuity, or on a medium that has different propagation characteristics, such as its dielectric constant. (2) In a transmission line, a current or voltage wave that travels in the direction from its source to the load.

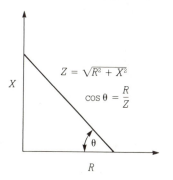

Impedance triangle

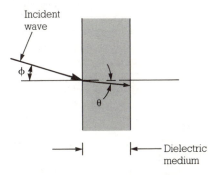

Incident wave impinging on a different medium

induced current A current produced in a conductor by a time-varying electromagnetic field.

induced voltage A voltage produced in a circuit by a change in the number of magnetic lines of force passing through a coil in the circuit.

inductance Symbolized by the letter L and measured in henries, the property of a circuit or circuit element that opposes a change in current flow, and causes the current through it to lag the voltage across it by 90 degrees.

induction (1) The process whereby a magnetic field or charge is produced as a result of the application of an electric or magnetic field. (2) The production of an electromotive force in a conductor either as a result of its motion through a magnetic field or as a result of a change in the magnetic flux surrounding it. See *Faraday's law of induction.*

induction density See *flux density.*

induction field (1) The electromagnetic field of a coil carrying alternating current, responsible for the voltage induced by that coil in itself or in a nearby coil. (2) In a transmitting antenna, the portion of the electromagnetic field that acts as if it were permanently associated with the antenna.

induction motor An AC motor in which the primary winding of the stator is connected to the power source, while a polyphase, or squirrel-cage secondary winding on the rotor carries the induced current.

inductive (1) Pertaining to inductance. (2) Pertaining to the induced voltage as a result of mutual inductance, or the induced electric charge by electrostatic induction.

inductor Symbolized by the letter L on schematic diagrams, a device designed to introduce inductance in a circuit; a coil.

infrared radiation Abbreviated *IR radiation,* electromagnetic radiation of wavelengths between 0.75 and 1 000 microns, which is between microwaves and visible red light. This radiation spectrum is further divided into near infrared (0.75–3 microns), middle infrared (3–30 microns) and far infrared (30–1 000 microns).

in phase (1) Two or more waveforms that have the same frequency and that pass through corresponding values of their cycle at the same time. (2) No phase shift.

input (1) The signal applied to a network or device. (2) The terminals through which an input is supplied.

input admittance (1) The reciprocal of input impedance, expressed in siemens. (2) The admittance between the input terminals of a four-port network when its two output terminals are shorted together.

input impedance The impedance a circuit or network presents to its source. See *driving point impedance.*

input offset voltage The voltage that must be applied to one of the input terminals of an operational amplifier to produce a zero output voltage. Also called *input voltage offset.*

input/output (1) Abbreviated *I/O,* the transmission of information from an external source to a computer or from a computer to an external source. (2) A general term applied to the equipment used in communicating with a computer.

input voltage offset See *input off-set voltage.*

insertion gain The gain expressed in decibels that is the ratio of the power delivered to a load following the introduction of a network, such as an amplifier in series with another system, to the power delivered to the load without the additional network.

insertion loss The loss expressed in decibels that is the ratio of the power delivered to a load following the introduction of a network, such as an attenuator in series with another system, to the power delivered to the load without the additional network.

instantaneous value The value of a varying quantity at a particular instant in time.

instrumentation amplifier Abbreviated *IA*, a closed-loop, high-gain, high-impedance, differential amplifier with a high common-mode rejection over a wide frequency range.

instrument multiplier A highly accurate resistor used in series with a voltmeter to extend its voltage range. Also called a *voltage-range multiplier.*

instrument shunt A highly accurate resistor used in parallel with an ammeter to extend its current range.

insulated gate field effect transistor Abbreviated *IGFET*, a field effect transistor whose gate is insulated from the channel, thus creating a higher input impedance than a junction field effect transistor. Also called a *MOSFET.*

insulator (1) A material having very high resistance and dielectric strength. (2) A structure made of such a material, used to support an energized conductor while keeping it electrically isolated.

integrated circuit Abbreviated *IC*, and designated by *U* on a schematic diagram, an electronic circuit in which all of the active and passive components are made from a single semiconductor substrate. Also called a monolithic integrated circuit.

integrating filter A filter in which successive voltage pulses cause accumulated buildup of charge and voltage on an output capacitor.

integrator (1) A circuit, either active or passive, whose output is the time integral of its input signal so that

$$V_{out} = \int_{t_1}^{t_2} v(t)\ dt$$

(2) A low-pass filter.

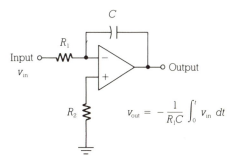

Operational amplifier integrator

intensity The strength, magnitude, or amount of a quantity, such as voltage.

intensity modulation See *Z-axis modulation.*

interbase resistance The resistance between both base leads in a unijunction transistor.

interchannel interference Interference produced in a common channel by signals from one or more other channels.

73

interconnected star connection
See *zig-zag connection.*

interference (1) Any unwanted energy or signal that interferes with the reception or display of desired signals. (2) Noise.

interlock A switch or other device that prevents the activation or operation of a circuit or equipment when a hazard exists.

interlock circuit A circuit in which one action cannot occur until one or more other actions have taken place.

interlocked relay See *bistable relay.*

interlock switch A safety switch that opens the path to current flow when a door or panel is opened or removed.

intermediate frequency Abbreviated *IF*, the sum or difference frequency produced by mixing an input frequency with that of a local oscillator.

intermittent Not continuous.

intermodulation Abbreviated *IM*, the modulation of frequency components of a complex waveform by each other, thus producing new waveforms whose frequencies are equal to the sums and differences of the integral multiples of the frequency components of the original complex waveform.

intermodulation distortion A type of nonlinear distortion characterized by the presence of output frequency components equal to the sums and differences of the frequency components of the input signal.

International System of Units
Abbreviated *SI*, a system of units adopted internationally in 1960 in which the ampere, candela, kelvin, kilogram, meter, mole, and second are the seven basic units from which all other units of measure are derived.

interpole An auxiliary pole placed between the main poles of a rotary DC machine in order to provide additional flux, assisting commutation.

intrinsic standoff ratio Symbolized by the Greek letter η, a dimensionless quantity of a unijunction transistor equal to the resistance of base 1 divided by the interbase resistance.

inverse feedback See *negative feedback.*

inverse peak voltage See *peak inverse voltage*

inverter (1) A circuit that converts DC to AC. (2) A circuit whose magnitude of output intensity is the same as the input, but whose polarity is changed.

inverting amplifier (1) An amplifier whose output signal is inverted, or 180 degrees out of phase with its input. (2) Grounded cathode, common emitter, or common-source amplifiers. (3) An operational amplifier circuit whose output is inverted with respect to its input, and whose voltage gain can be less than, equal to, or greater than 1.

inverting input One of two inputs of an operational amplifier or comparator.

I/O Abbreviation for *input/output.*

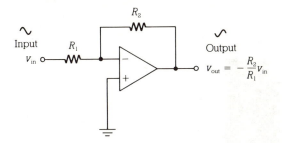

$$V_{out} = -\frac{R_2}{R_1}V_{in}$$

Inverting amplifier

ion   A charged particle formed when one or more electrons are taken from or added to a neutral atom or molecule.

ips   Abbreviation for *inches per second.*

IR   Abbreviation for *infrared.*

IR drop   From Ohm's law, the voltage developed across a resistance R when there is a current I passing through it.

iron-core coil   A transformer having iron laminations that make up all or part of the path for magnetic lines of force that link the transformer windings.

iron loss   See *core loss.*

IR radiation   Abbreviation for *infrared radiation.*

isolation amplifier   (1) A unity gain amplifier that provides total isolation between input and output signals. (2) A buffer.

isolation transformer   A transformer used to separate one section of a system from undesired influences of other sections. It usually has a 1:1 turns ratio to eliminate a direct connection without changing voltages.

isotropic   Having identical properties in all directions.

j A complex operator that is equal to the square root of − 1.

J (1) Unit symbol for *joule*. (2) Symbol for *electric current density*. (3) Letter designation for input or output *jacks* and *sockets* on schematic diagrams.

jack Designated by the letter *J* on schematic diagrams, a connecting device to which a wire or group of wires of a circuit may be attached and electrically connected to other circuits by means of insertion of a corresponding mating plug. Also called a *socket*.

jam To electronically interfere with the reception of radio signals.

JAN specification Acronym for Joint Army-Navy specification, the forerunner of the present Military (MIL) specifications.

J-display A radar display in which the time base is a circle and targets appear as radial deflections from the time base.

JEDEC Acronym for Joint Electron Device Engineering Council.

JETEC Acronym for Joint Electron Tube Engineering Council.

JFET Acronym for *junction field effect transistor*.

jitter Small, rapid variations in a waveform caused either by mechanical vibration, changes in supply voltage, or system instability. Also called *flicker*.

JK flip-flop A clocked flip-flop comprised of a latch and steering gates that changes the state of its output if both the *J* and *K* inputs are at logic 1 (high) when a clock pulse arrives.

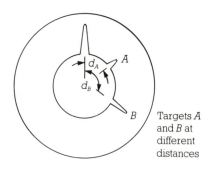

Targets *A* and *B* at different distances

J-display

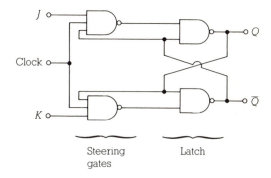

Steering   Latch
gates

JK flip-flop

Johnson noise   The noise power in watts generated by a resistor at a temperature above absolute zero, according to the formula $N = KTB$, where N = noise power in watts, K = Boltzmann's constant, 1.38047×10^{-23} joules per degree Kelvin, T = absolute temperature in degrees Kelvin, and B = signal bandwidth in hertz. Also called *resistance noise,* or *thermal noise.*

joint   A common connection between two or more conductors.

joule   Abbreviated J, the unit of work or energy. The work done by a force of 1 newton acting on a distance of 1 meter. Equal to 1 watt-second, or the work done in one second in maintaining a current of 1 ampere flowing in a resistance of 1 ohm. The energy required to move a charge of 1 coulomb between two points having a potential difference of 1 volt. Equal to 10^7 ergs.

Joule, James Prescott   An English physicist (1818–1889) who first determined the mechanical equivalent of heat, and after whom the electrical unit of work was named.

Joule effect   The effect whereby electrical energy is converted into heat by current flowing in a conduc-

tor as a consequence of its resistance by an amount equal to I^2R. Half of this heat flows to the hot junction and the other half flows to the cold junction.

Joule heat   The thermal energy, or heating, resulting from the Joule effect.

Joule's law   The rate at which heat is produced in a constant-resistance electric circuit is proportional to the square of the current.

J/S ratio   Expressed in decibels, the ratio of the total interference power to the carrier signal power in the transmission medium at the receiver.

juice   Slang term for electric current.

jumper   A length of conductor used to make an electrical connection between two points or terminals of a circuit, or to provide a path around a break in the circuit.

junction diode   A two-terminal device consisting of both p-type (anode) and n-type (cathode) semiconductor materials joined to form a junction; it is a basic building block of the transistor. As a diode, it conducts current easily from its anode to its cathode, but does not conduct current from cathode to anode.

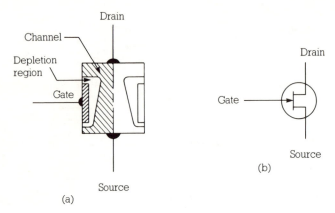

Junction field effect transistor

junction field effect transistor
Abbreviated *JFET*, a transistor consisting of a gate region which is diffused into a channel region. When a voltage is applied to the gate with respect to the JFET's source lead, the channel is either depleted (negative polarity) or enhanced (positive polarity), so that the drain current is controlled. Also called a *unipolar transistor,* or *field effect transistor.*

junction transistor   A transistor having three alternate sections of p-type and n-type materials, so that the sections are in either a p-n-p or an n-p-n arrangement.

k Prefix symbol for *kilo,* 10^3.

K (1) Symbol for *cathode, gage factor.* (2) Unit symbol for *kelvin.* (3) Letter designation for *relay* or *solenoid* on schematic diagrams.

kappa (κ) Tenth letter of the Greek alphabet, used to represent *coupling coefficient,* and *susceptibility.*

K-band The radio-frequency band from 10.9 to 36 GHz.

kc Abbreviation for kilocycle, now replaced by kHz.

kelvin Abbreviated *K,* a unit of temperature used without the degree symbol (°). Formerly degree Kelvin.

Kelvin, Baron William Thompson A British mathematician and physicist (1824–1907) who introduced the absolute, or kelvin, temperature scale that bears his name.

Kelvin balance An instrument for measuring current achieved by passing the unknown current through both a fixed and a movable coil attached at one arm of a balance. The resultant force between the coils is then compared with the force of gravity acting on a known weight at the other end of the balance arm.

Kelvin bridge Used to measure very low resistances. An adaptation of the Wheatstone bridge, the Kelvin bridge has seven arms, offsetting the parasitic resistance of its leads and contacts. Also called a *double,* or *Thompson bridge. (See figure on page 80.)*

Kelvin scale A temperature scale using the same divisions as the Celsius scale, but with the zero point at absolute zero, or -273.15°C. Also called the *absolute temperature scale.*

kHz Unit symbol for kilohertz.

kickback (1) The electromotive force developed across an inductor when the current through it is reduced abruptly. (2) A countervoltage.

kilo Abbreviated *k,* prefix for 10^3.

Kirchhoff, Gustav Robert A German physicist (1824–1887) who, with Bunsen, discovered the elements cesium and rubidium and also did research in electricity and thermodynamics.

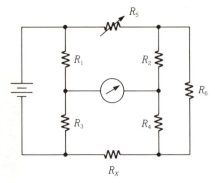

$$R_X = \left(\frac{R_3}{R_1}\right) R_5 + R_6 \left(\frac{R_4}{R_2 + R_4 + R_6}\right) \left(\frac{R_3}{R_1} - \frac{R_4}{R_2}\right)$$

Kelvin bridge

Kirchhoff's current law   The sum of the currents entering a node or junction equals the sum of the currents leaving the node or junction.

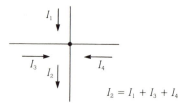

$$I_2 = I_1 + I_3 + I_4$$

Kirchhoff's current law

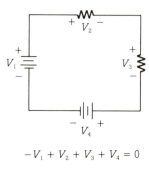

$$-V_1 + V_2 + V_3 + V_4 = 0$$

Kirchhoff's voltage law

Kirchhoff's radiation law   At any given temperature and for the same kind of radiation, the absorptivity and emissivity of a body are equal.

Kirchhoff's voltage law   The algebraic sum of the voltage drops across the branches of any closed path or loop is zero.

knee   An abrupt change in direction between two relatively straight segments of a curve.

knife switch   A switch having a knifelike blade or wiper which can be pushed between spring contacts.

knockout   A section of a metal cabinet, chassis, or outlet box which can be easily removed for the attachment of cables, fittings, etc.

kW   Unit symbol for kilowatt.

80

L (1) Mathematical symbol for *inductance.* (2) Unit symbol for *lambert.* (3) Letter designation for an *inductor* or *coil* on schematic diagrams.

ladder network A network composed of H-, L-, T-, or pi-networks connected in series. Also called a *series-shunt network.*

lag (1) The opposite of lead. (2) The difference, expressed in radians or electrical degrees, between two events or values.

lag network (1) A circuit whose output signal lags the input signal. (2) A low-pass filter, either active or passive.

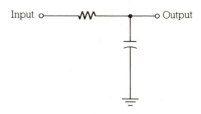

Lag network

lambda (Λ, or λ) Eleventh letter of the Greek alphabet, used to designate *wavelength.*

lambda diode A two-terminal solid-state negative resistance device that is essentially the equivalent of an interconnected pair of complementary field-effect transistors. *(See figure on page 82.)*

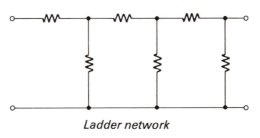

Ladder network

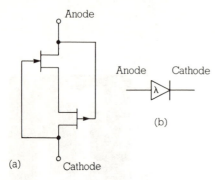

Lambda diode: (a) Equivalent circuit; (b) Symbol

lambert Symbolized by L, a unit of luminance. Equal to $1/\pi$ candela per square centimeter.

Lambert, Johann Heinrich A German-French philosopher and scientist (1728–1777) who did research on measurement of light intensity and degrees of heat and humidity, and after whom the unit of light intensity was named.

Lambert's law of illumination
The illumination of a surface on which the light falls perpendicular from a point source is inversely proportional to the square of the distance of the surface to the source. If the line perpendicular to the surface makes an angle with the direction of the rays, the illumination is proportional to the cosine of that angle.

lamp (1) A device that produces light. (2) A light bulb.

lamp cord A pair of parallel wires, usually No. 18 or No. 20, used primarily for connecting electric equipment to wall outlets.

Laplace, Pierre Simon A French astronomer and mathematician (1749–1827) who, with J. L. Lagrange, established Newton's hypothesis of gravitation.

Laplace's law The magnetic field strength at any given point, as a result of an element of a current-carrying conductor, is directly proportional to the strength of the current and the projected length of the element, and is inversely proportional to the square of the distance of the element from the given point.

Laplace transform A special case of the Fourier transform (one-sided) converting a time-dependent function into the complex frequency domain by performing the operation

$$F(s) = \int_0^\infty f(t)e^{-st}\, dt$$

where $s = \sigma + j\omega$.

large signal Pertaining to an input signal that is large enough so that any active element behaves in a nonlinear fashion.

LASCR Abbreviation for *light-activated silicon controlled rectifier*.

LASCS Abbreviation for *light-activated silicon controlled switch*.

laser (1) Acronym for light amplification by stimulated emission of radiation. (2) An active electron device that converts input power into a very narrow, intense beam of coherent visible or infrared light. The input power excites the atoms of an optical resonator to a higher energy level and the resonator forces the excited atoms to radiate in phase. The four major laser types are gas, liquid, semiconductor, and solid.

latching relay A relay whose contacts lock in the energized or de-energized position until reset either manually or electrically. Also called a *bistable relay*.

lattice The geometrical arrangement of atoms in a crystal.

lattice network A balanced network of four branches connected in a loop where two nonadjacent junction points serve as input terminals, while the remaining two junction points serve as output terminals.

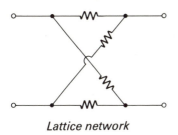

Lattice network

law of electric charges Like charges repel; unlike charges attract.

law of electromagnetic induction See *Faraday's law of induction.*

law of electrostatic attraction See *Coulomb's law.*

law of induced current See *Lenz's law.*

law of magnetism Like poles repel; unlike poles attract.

L-band The radio-frequency band from 390 MHz to 1.55 GHz.

LCD Abbreviation for *liquid-crystal display.*

LC-network Any parallel or series circuit made from a capacitor and an inductor.

L-display A modified radarscope A-display in which a target appears as a pair of vertical deflections. When the radar antenna is correctly pointed at the target in azimuth, the deflections are of equal height. When not correctly pointed, the difference

in pulse heights gives an indication of the magnitude and direction of azimuth pointing error.

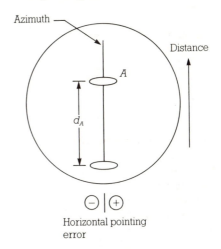

L-display

lead (1) A wire that connects two points in a circuit. (2) The opposite of lag. (3) The difference, expressed in radians or electrical degrees, between two events or values.

lead-acid cell A cell in a storage battery which consists of platelike, lead electrodes surrounded by a dilute sulfuric acid electrolyte. Also called a lead cell.

leading edge The major portion of the rise of a pulse waveform.

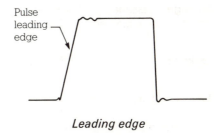

Leading edge

lead network (1) A circuit whose output signal leads the input signal. (2) A high-pass filter, either active or passive.

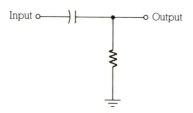

Lead network

leakage (1) Electrical loss from poor insulation. (2) Unwanted current flow through an insulator, or through certain components such as capacitors.

leakage current An unwanted, but small, flow of current.

leakage flux The magnetic lines of force that extend beyond an intended area.

leakage inductance The self-inductance caused by the leakage flux produced in a transformer's windings.

leakage reactance The inductive reactance produced by magnetic flux in one coil of a transformer, which couples with the other coil of the transformer.

leakage resistance The resistance path through which leakage current flows.

leakance The reciprocal of insulation resistance.

leaky A slang term usually applied to a capacitor whose internal resistance has decreased down to a point so that a DC current flows.

LED Acronym for *light-emitting diode.*

LED display A visual, alphanumeric display using a series of light-emitting diodes in a single, monolithic package, configured either in common-anode or common-cathode arrangements.

left-hand rule See *Fleming's rules.*

Lenz's law If a current is induced in a circuit as the result of a magnetic field or a change in magnetic flux, the direction of the induced current is such as to exert a magnetic force opposing the motion or flux change. Also called the *law of induced current.*

level (1) The magnitude of a measured quantity in relation to an arbitrary reference value expressed in like units. (2) The ratio of the measured value to a reference value (e.g., decibels).

leyden jar The earliest construction of a capacitor where a glass jar was lined on both the inside and outside walls (the dielectric) with metal foil (the parallel plates).

LF (or lf) Abbreviation for *low frequency.*

light Energy in the form of electromagnetic radiation visible to the human eye, between 4 000 and 7 700 angstroms, although it sometimes is used to also include infrared and ultraviolet radiation.

light-activated silicon controlled rectifier Abbreviated *LASCR*, a silicon controlled rectifier which is triggered by the application of light.

light-activated silicon controlled switch Abbreviated *LASCS*, a solid-state diode which conducts upon the application of light.

light-emitting diode Abbreviated *LED*, a semiconductor diode which emits light when forward biased.

light relay A relay circuit whose contacts close when the intensity of a light beam changes.

limit bridge A form of the Wheatstone bridge which is used for rapid routine testing during the production process. Its use is primarily concerned with resistance tolerance limits, rather than exact values.

limiter A circuit that prevents the amplitude of a waveform from exceeding a given level. A limiter may act on positive peaks, negative peaks, or both.

limit ratio The ratio of peak value to limited value.

line (1) A conductor of electricity or energy, such as a wire or transmission line. (2) A single trace of an electron beam across a screen in a cathode ray tube. (3) Another term for *maxwell,* equal to 10^{-8} weber.

linear (1) A type of input-output relationship whose output is directly proportional to a given input. (2) A straight-line relationship between two variables.

linear amplifier An amplifier whose output waveform, with no resulting distortion, is directly proportional to its input.

linear differential transformer A transformerlike electromechanical transducer that converts motion or displacement into an output voltage whose amplitude and phase are pro-

portional to position. Sometimes referred to as a *linear motion transducer.*

linear integrated circuit (1) A class of integrated circuits whose output is linearly related to the input. (2) A term frequently applied to analog integrated circuits such as operational amplifiers, timers, and voltage regulators.

linear motion transducer An electromagnetic component that translates linear mechanical motion or position into an AC output voltage. When a movable magnetic core is displaced axially in one direction from its resting position, the output voltage is in phase with the excitation voltage; in the opposite direction, it is 180 degrees out of phase. At rest, the output is zero. Also called *linear differential transformer,* or a *linear variable differential transformer.*

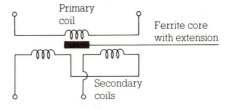

Linear motion transducer

linear taper A type of potentiometer whose resistance changes linearly as it is varied throughout its resistance range.

linear variable differential transformer Abbreviated *LVDT*, an electromechanical transducer which translates displacement or position changes into AC voltage changes. Also called a linear motion transducer, or a *differential transformer. (See figure on page 86.)*

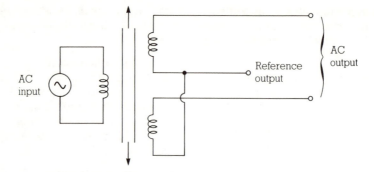

Linear variable differential transformer

line balance (1) The degree to which the conductors of a transmission line are electrically alike to each other and ground. (2) The impedance equal to that of the line at all frequencies.

line cord A two- or three-wire cord that is used to connect any device to an AC power line. Also called a *power cord.*

line drop A voltage loss that occurs between two points in a power or transmission line due to either the resistance, reactance, or leakage of the line.

line equalizer A reactive circuit inserted in a transmission line to change its frequency response characteristic.

line filter A circuit, generally passive, which is inserted between a transmitter or receiver and the antenna, or between an appliance and the power line to block the transmission or passage of unwanted signals or noise.

line frequency (1) In television, also called the horizontal line frequency or horizontal frequency, the number of times per second the scanning spot crosses a fixed vertical line in the picture in one direction including vertical return intervals. (2) The frequency of the AC power line, generally 60 Hz in the United States and 50 Hz in the rest of the world.

line loss The total energy loss in a transmission line.

line noise Noise originating in a transmission line. See *transmission loss.*

line-of-sight distance (1) The distance to the earth's horizon from an elevated distance. (2) The maximum unobstructed optical or visual path between two points on the earth's surface.

line regulation The change in output, usually voltage, or a regulated power supply for a specified change in line voltage, usually expressed as a percentage.

liquid crystal display Abbreviated *LCD,* a visual, alphanumeric display constructed from applying a current across two glass plates with a nematic liquid crystal solution between them.

Lissajous figures A pattern produced when sinewave signals of various amplitude, phase, and frequency relationships are applied simultaneously to the horizontal and vertical amplifiers of an oscilloscope.

litz wire A wire formed of very fine separately insulated interwoven strands to reduce skin effect, or lower resistance to high-frequency circuits. Also called Litzendraht wire.

live A slang term applied to a circuit that has current flowing through it, such as a wire.

lm Unit symbol for *lumen,* the unit of luminous flux.

L-network A passive network composed of two impedance branches in series.

load (1) The amount of electrical power drawn from a particular

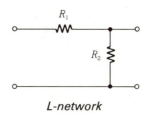

L-network

source. (2) A device that absorbs electric power. (3) The impedance to which energy is applied.

loaded-Q The quality factor of an impedance when coupled or connected under operating conditions. Also referred to as *working-Q.*

loading coil An inductor inserted in a circuit, primarily an antenna, to increase its inductance, which in turn increases the overall length of an antenna.

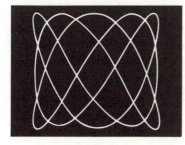

3:2 ratio

3:2 ratio

5:3 ratio

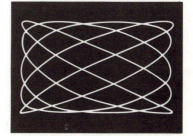

4:5 ratio

Lissajous figures

87

loading error   The error as a result of the loading effect of a load on a signal source driving it.

load line   A line drawn across a set of characteristic curves of a nonlinear device, such as a bipolar transistor, so that it defines the combination of all possible operating points for a given load condition.

load matching   Adjusting the load impedance to produce the desired energy transfer from the source to the load.

load saturation curve   The saturation curve of a machine for a specified load current.

lobe   In an antenna, a portion of the directional pattern bounded by one of two cones of nulls.

local oscillator   An oscillator in a mixer circuit whose output is combined with another signal to produce a sum or difference frequency, called the intermediate frequency.

lock   (1) To synchronize or become synchronized with. (2) To follow or control.

locked relay   See *bistable relay*.

lockout relay   An electrical or manual reset relay whose function is to maintain circuits inoperative until it is reset.

lock range   In a phase-locked loop, the frequency range over which the output follows changes in the input.

logarithm   Sometimes called log, a number y related to another number

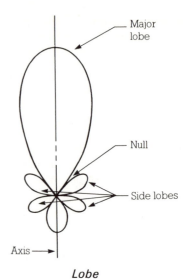

Lobe

z by the relation $x^y = z$, where x is any number greater than 1. Logarithms have the basic properties such that $\log(AB) = \log(A) + \log(B)$ and $\log A^B = B \log(A)$. The two most frequently used systems are the common logarithms (base 10, $x = 10$) and the natural logarithms (base e, $x = 2.71828$). The common logarithm of the number A is written $\log_{10} A$ or simply $\log A$; the natural logarithm of the number A is written $\log_e A$ or simply $\ln A$.

logarithmic amplifier   An amplifier whose output is proportional to the logarithm of its input.

logarithmic decrement   See *damping constant*.

logarithmic scale   A scale on which the various points are plotted according to the logarithm of the number with which the point is labeled.

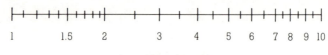

Logarithmic scale

log-periodic antenna A class of broadband antennas whose electrical properties are periodic with the natural logarithm (i.e., base e) of frequency.

long wave A radio frequency wave having a wavelength generally longer than 1 000 meters, corresponding to frequencies less than 300 kHz.

loop In an electrical network, a closed path of two or more branches, or circuit elements. Also called a *mesh*.

loop gain The gain of an amplifier when the amplifier's feedback path is removed.

loose coupling A degree of coupling less than critical for two resonant systems.

LORAN An acronym for long range navigation, in which radio signals between 1.8 and 2 MHz from three fixed stations are used to determine one's position.

Lorentz, Hendrick Antoon A Dutch Nobel prize physicist (1853–1928) who pioneered in formulating the relations between electricity, magnetism, and light.

Lorentz force The force F exerted on an electric charge q moving with a velocity v as a result of both an electric field E and a magnetic field so that $F = q(E + vB \sin \theta)$.

loss (1) A reduction of power of a signal as it is transmitted from one point to another, usually expressed in decibels. (2) Energy dissipated without associated useful work.

lossless line A theoretically perfect transmission line, i.e., one that has no losses so that it transmits all of the energy fed to it.

loudspeaker See *speaker.*

low band Generally the frequency range occupied by television channels 2 through 6, or 54–88 MHz.

lower sideband Abbreviated *LSB*, the frequency or group of frequencies generated below the carrier frequency in an amplitude modulated system.

low frequency Abbreviated *LF*, any frequency in the region from 30 to 300 kHz.

low-pass filter (1) A circuit which passes all frequencies without attenuation below a specified frequency while rejecting all frequencies above the specified frequency. (2) A lag network. *(See figure on page 90.)*

L-pad A form of an L-network that has practically the same impedance at all settings when both impedance branches are adjusted simultaneously.

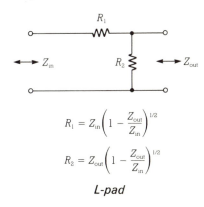

$$R_1 = Z_{in}\left(1 - \frac{Z_{out}}{Z_{in}}\right)^{1/2}$$

$$R_2 = Z_{out}\left(1 - \frac{Z_{out}}{Z_{in}}\right)^{1/2}$$

L-pad

LSB Abbreviation for *lower sideband.*

lug A device crimped or soldered at one end of a wire lead that can be used to easily connect the wire lead to a binding screw or post.

lumen

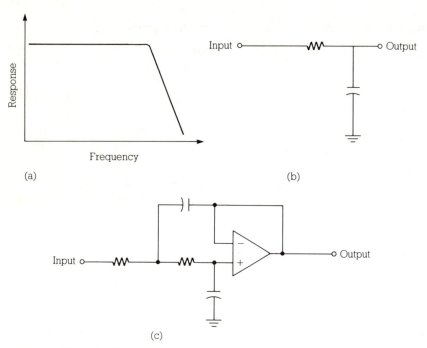

Low-pass filter: (a) Frequency response; (b) Passive circuit; (c) Active circuit

lumen   Abbreviated *lm*, a unit of luminous flux equal to the quantity of light through a solid angle of 1 steradian from a uniform point light source of 1 candela.

luminance   The luminous intensity of a surface in a given direction. Measured either in lamberts, or candela per square centimeter.

lumped   Concentrated in single, discrete circuit elements rather than being distributed throughout a system.

lux   Abbreviated *lx*, a unit of measure of illumination. Equal to 0.0929 foot-candle, or 1 lumen per square meter.

LVDT   Abbreviation for *linear variable differential transformer.*

lx   Unit symbol for *lux,* the unit of illuminance.

m (1) Mathematical symbol for *modulation coefficient.* (2) Unit symbol for *meter.* (3) Prefix symbol for *milli* (10^{-3}).

M (1) Mathematical symbol for *mutual inductance, magnetic field strength, magnetization.* (2) Prefix symbol for *mega* (10^6). (3) Letter designation for a *meter* on schematic diagrams.

magnet An object that produces a magnetic field surrounding itself, so that it attracts or repels other magnetic objects.

magnetic circuit A completely closed path for magnetic lines of force, having a reluctance that limits the magnetic flux which can be sent through the circuit by the magneto-motive force.

magnetic core A quantity of ferrous material placed in a coil or transformer to provide a better path than air for magnetic flux. As a result, the inductance of the coil is increased as well as the coupling between the windings of a transformer. Also called a *core.*

magnetic field Any space or region in which a magnetic force is exerted on moving electric charges, which may be produced by current flowing in a coil or wire, a magnet, or the earth itself.

magnetic field intensity See *magnetic field strength.*

magnetic field strength Symbolized by the letters H or M, and measured in amperes per meter, the magnitude and direction of the magnetic field vector. Also called *magnetizing force* or *magnetic field intensity.*

magnetic flux Symbolized by the Greek letter Φ and measured in webers, the number of lines of force that pass through a given surface in a magnetic field.

magnetic flux density Symbolized by the letter B and measured in teslas, the number of magnetic lines of force per unit area perpendicular to force lines. Also called *magnetic induction.*

magnetic induction See *magnetic flux density.*

magnetic line of force (1) An imaginary line whose tangent at any point represents the direction of magnetic flux. (2) A unit of magnetic flux equal to 1 maxwell.

magnetic pole (1) One of two poles of a magnet in which the magnetic intensity is greatest, known as the north and south poles. (2) Either of two locations on the earth's surface toward which a compass needle points.

magnetic reed relay See *reed relay*.

magnetism (1) The property exhibited by a magnet or by a ferromagnetic material surrounded by an electric current so that it can produce or conduct magnetic lines of force that are capable of interacting with electric fields or other magnetic fields. (2) The science of the study of magnetic phenomena.

magnetization (1) Symbolized by *M*, the degree to which a particular object is magnetized. (2) The process of magnetizing a magnetic material.

magnetization curve See *B-H curve*.

magnetizing force See *magnetic field strength*.

magnetoelectric Pertaining to the generation of voltages as a result of magnetic fields, such as in a generator.

magnetohydrodynamics Abbreviated *MHD*, the study of the effects of magnetic fields on superheated ionized gases and conducting fluids.

magnetomotive force Abbreviated *MMF* and symbolized by F_m, the force that produces a magnetic field. It is the total magnetizing force acting around a completely closed magnetic circuit and corresponds to the voltage in an electric circuit. If a current is flowing in a coil of wire, the magnetomotive force is proportional to the number of ampere-turns, or gilberts.

main A line that brings power from a generator, converter, or service-cutoff switch to the main distribution center, or service panel for power lines inside a building.

mains British term for *power line*.

majority carrier In a semiconductor, current carriers that are in the majority. In a p-type semiconductor majority carriers are holes; in an n-type semiconductor, they are electrons.

make-before-break switch A type of switch having a double-throw contact which switches on one circuit before breaking the other circuit.

make contact A normally open stationary relay contact, closed when the relay is energized.

male connector A connector with one or more protruding pins, usually a plug, which mates with a corresponding female connector. See *plug*.

manganin An alloy that contains 84 percent copper, 12 percent manganese, and 6 percent nickel, used in the manufacture of precision wire-wound resistors because of its low temperature coefficient of resistance.

Marconi, Gugliemo Marchese An Italian physicist (1874–1937) who developed wireless telegraphy and shared the Nobel prize in physics in 1909. Also, in the field of electromagnetic waves, he improved upon the inventions of others and developed a type of antenna, known as the Marconi antenna.

Marconi antenna A type of antenna that is connected to ground at one end through the receiver or transmitter input coil and suitable matching circuits.

master oscillator An oscillator that generates the carrier frequency of the output of an amplifier or transmitter.

matched load A load whose impedance equals that of the driving source, which results in maximum power transfer from the source to the load.

matching The process of connecting two circuits or devices together so that their impedances are equal or are equalized by a coupling device, such as a transformer, to give a maximum transfer of energy.

maximally flat filter See *Butterworth filter.*

maxwell Abbreviated *Mx*, the unit of magnetic flux. Equal to 1 gilbert per square centimeter, or 10^{-8} weber. Also called *line.*

Maxwell, James Clerk A Scottish physicist (1831–1879) who advanced the mathematics of the electromagnetic field. He concluded that electricity and magnetic energy travel in transverse waves and that light waves exhibit similar behavior. After whom the unit of magnetic flux was named.

Maxwell bridge An AC bridge that measures inductance or capacitance and whose bridge balance depends on frequency. Also called a *Maxwell-Wien bridge.*

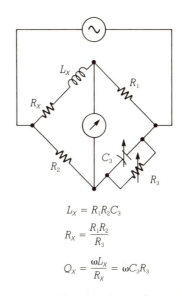

$$L_X = R_1 R_2 C_3$$

$$R_X = \frac{R_1 R_2}{R_3}$$

$$Q_X = \frac{\omega L_X}{R_X} = \omega C_3 R_3$$

Maxwell (Maxwell-Wien) bridge

Maxwell inductance bridge An AC bridge which compares the ratio of inductances in two adjacent arms to the ratio of two resistors in the other two arms. Bridge balance depends on frequency. Also called a Maxwell inductance comparison.

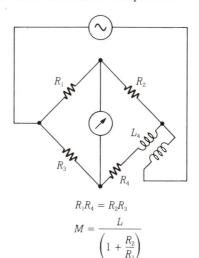

$$R_1 R_4 = R_2 R_3$$

$$M = \frac{L}{\left(1 + \frac{R_2}{R_1}\right)}$$

Maxwell inductance bridge

93

Maxwell's equations A set of four equations developed by James C. Maxwell that form the basis for electromagnetic theory. In terms of the magnetic field strength (H), magnetic flux density (B), electric field strength (E), electric charge density (ρ), current density (J), and time (t), the equations are expressed in derivative form as

$$\nabla \times \mathbf{H} = \frac{\partial \mathbf{D}}{\partial t} + \mathbf{J}$$

$$\nabla \times \mathbf{E} = -\frac{\partial \mathbf{B}}{\partial t}$$

$$\nabla \cdot \mathbf{B} = O$$

$$\nabla \cdot \mathbf{D} = \rho$$

Maxwell-Wein bridge See *Maxwell bridge.*

m-derived filter section A T- or pi-network section designed so that when two or more sections are cascaded, their impedances are matched at all frequencies, even though the sections may have different resonant frequencies.

measurand A measured physical quantity.

medium frequency Abbreviated *MF*, any frequency in the region from 300 kHz to 3 MHz.

mega Abbreviated *M*, prefix for 10^6.

Megger Trademark name of the James C. Biddle Company for a high-range ohmmeter using a hand-driven DC generator as its voltage source.

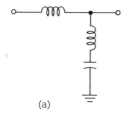

(a)

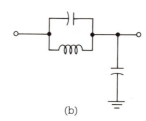

(b)

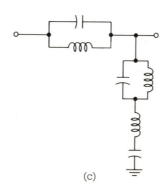

(c)

m-derived filter sections: (a) Lowpass, series type; (b) Lowpass, shunt type; (c) Notch, series type

Tilted   Level

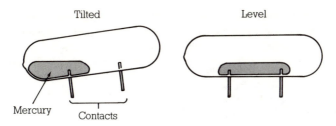

Mercury   Contacts

Mercury switch

mercury arc tube A discharge or rectifier tube filled with mercury vapor.

mercury cell A cell having a constant output voltage by a chemical reaction between zinc and mercury oxide.

mercury switch A switch that is closed by making a region of mercury move up to a pair of contacts and bridging them.

mercury vapor rectifier A rectifier tube filled with mercury vapor instead of a vacuum.

mesh A set of branches that forms a closed path in such a way that if one branch is omitted from the set, the remaining branches do not form a closed loop. Also called a *loop*.

metal oxide semiconductor
Abbreviated *MOS*, a metal-insulator-semiconductor structure in which the insulating layer is an oxide of the substrate material, such as silicon dioxide.

metal oxide semiconductor field effect transistor Abbreviated *MOSFET*, a field effect transistor in which the gate is insulated from the channel by a metal-oxide dielectric film. Consequently, this allows for a higher input impedance by making the gate forward biased to enhance the conductivity of the channel. Also

called an *insulated gate field effect transistor,* or *IGFET.*

meter (1) Abbreviated *m*, the unit of length defined as equal to 1 553 164.13 wavelengths of the red light of a cadmium emission spectrum in air at 730 millimeters pressure at 0°C, and equal to 39.37 inches. (2) An indicating device that measures the value of a quantity, such as voltage or current, designated by the letter *M* on schematic diagrams.

MF Abbreviation for *medium frequency.*

MFD (or mfd) Outmoded abbreviation for microfarad, now replaced by μF.

MHD Abbreviation for *magneto-hydrodynamics.*

mho *Ohm* spelled backwards, the former unit of conductance, now replaced by *siemens.*

mic Abbreviation for *microphone.*

mica A transparent mineral that splits readily into thin sheets having excellent insulating properties. It is used as a dielectric in mica capacitors and as electrode spacers in vacuum tubes.

micro Symbolized by the Greek letter μ, prefix for 10^{-6}.

microcircuit An integrated circuit.

micron   Abbreviated by the Greek letter μ, a unit of measure equal to one millionth of a meter.

microphone   Abbreviated *mic*, an electroacoustic transducer that converts sound waves into equivalent electrical signals. Also called a *mike*.

microwave   Pertaining to wavelengths ranging from 0.3 to 30 cm, corresponding to frequencies from 1 to 100 GHz.

midband frequency   See *center frequency*.

midrange   Any audio frequency between 600 Hz and 6 kHz.

mike   Slang term for *microphone*.

Miller bridge   An AC bridge circuit that measures the amplification factor or transconductance of vacuum tubes.

Miller effect   The effect whereby the input capacitance of a triode vacuum tube amplifier is effectively the sum of the grid-cathode capacitance plus the product of the grid-plate capacitance and one plus the voltage gain of the triode stage, so that

$$C_{in}(\text{Miller}) = C_{gk} + (1 + A_v)C_{gp}$$

The increase of grid-cathode capacitance is due to the charge induced electrostatically on the grid by the anode through the grid-plate capacitance. To reduce this effect, a tetrode or pentode tube is used instead.

Miller integrator   An integrator circuit consisting of an operational amplifier with a series resistor and a capacitor in its feedback path. Also called a *Miller time base*.

Miller's theorem   A network transform that states that an amplifier circuit with a feedback impedance network connected from its output terminal to its input terminal may be replaced by an equivalent network having transformed impedances between its input and output terminals to ground, both of which depend on the amplifier's voltage gain and the feedback impedance. In addition, the feedback element is removed. The transformed input and output shunt impedances are given by

$$Z_{in} = \frac{Z_f}{A_v + 1}$$

$$Z_{out} = \frac{A_v Z_f}{A_v + 1}$$

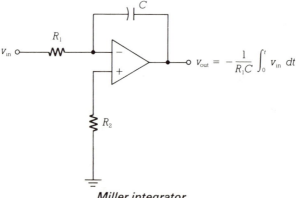

Miller integrator

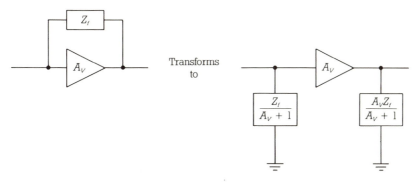

Miller's theorem

Miller time base See *Miller integrator.*

milli Abbreviated m, prefix for 10^{-3}.

minority carrier In a semiconductor, current carriers that are in the minority. In a p-type semiconductor, minority carriers are electrons; in an n-type semiconductor, they are holes.

mismatch factor See *reflection coefficient.*

mix See *heterodyne.*

MMF (or mmf) (1) Abbreviation for *magnetomotive force.* (2) Outmoded abbreviation for micro-microfarad, now replaced by picofarad.

modulation The process by which a characteristic of one signal is varied in proportion to the information contained in another signal. Types include amplitude, frequency, phase, pulse-amplitude, pulse-code, pulse-width, pulse-frequency, pulse-position, and pulse-time modulation.

modulation coefficient (1) Symbolized by the letter m and usually expressed as a percentage, the ratio of the peak variation in the modulation actually used in the transmitter to the maximum variation allowed without distortion. (2) In amplitude modulated systems, equal to the ratio of half the difference between the maximum and minimum amplitudes of the modulated wave to the average amplitude. (3) In frequency modulated systems, equal to the ratio of the instantaneous frequency deviation to the maximum frequency deviation allowed without distortion. Also called *modulation factor,* and *percent modulation.*

modulation envelope An imaginary curve connecting the peaks of a modulated waveform in order to more clearly show its shape. The envelope then represents the waveform of the intelligence carried by the signal. *(See figure on page 98.)*

modulation factor See *modulation coefficient.*

modulation index The ratio of the frequency deviation of a frequency modulation system to the frequency of the modulated wave when a sinewave signal is used.

modulator A transmitter circuit or device that varies the amplitude, frequency, phase, or other characteristic of a carrier signal in proportion to the waveform of the modulating signal that contains useful information.

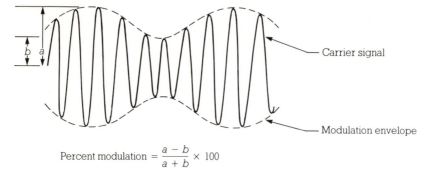

$$\text{Percent modulation} = \frac{a - b}{a + b} \times 100$$

Modulation envelope

monostable Having only one stable output state. See *monostable multivibrator.*

monostable multivibrator A multivibrator circuit whose output has only one stable state. When externally triggered, the output goes to its unstable state for a predetermined amount of time, called the *time delay* or *output pulse width,* and then returns to its stable state. The time that the output remains in its unstable state is generally controlled by a resistor-capacitor time constant. Also called a *monostable,* or *one-shot.*

MOS Abbreviation for *metal oxide semiconductor.*

MOSFET Abbreviation for *metal oxide semiconductor field effect transistor.*

motor A rotating machine that converts electrical energy into mechanical energy by means of the forces exerted by current-induced magnetic fields.

motorboating Oscillation, characterized by a series of pulses at very low audio frequencies, caused by excessive positive feedback. The result is a "putt-putt" sound similar to that produced by a motorboat.

mu (μ) Twelfth letter of the Greek alphabet, used to designate the *amplification factor* of a vacuum tube, or the prefix symbol for *micro.* Also the symbol for *micron.*

MUF Abbreviation for maximum usable frequency.

multimeter A meter that has a variety of measuring functions, such as the ability to measure current, voltage, and resistance. Also called a *volt-ohm-millammeter,* or *VOM.*

multiplex To combine two or more signals so they can be transmitted over a single channel.

multiplier (1) A circuit or device whose output is proportional to the product of both its inputs. (2) A device whose output frequency is a multiple of its input frequency. (3) In a voltmeter, a resistor used in series to increase its full-scale range.

Murray loop test A test method of locating a fault in a cable or transmission line, such as an accidental short to ground, by connecting the line to be tested with a good cable at two adjacent arms of a Wheatstone bridge.

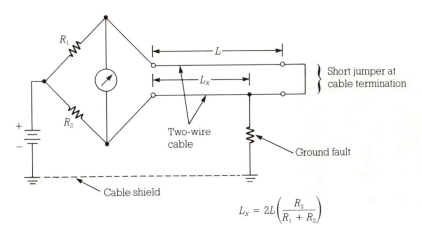

$$L_x = 2L\left(\frac{R_2}{R_1 + R_2}\right)$$

Murray loop test

mutual conductance See *trans-conductance*.

mutual inductance Symbolized by the letter *M*, a measure of the amount of inductive coupling existing be-

tween two coils, related to the flux linkages produced in one coil by the current in the other coil.

Mx Unit symbol for *maxwell*.

n Prefix symbol for *nano,* 10^{-9}.

N (1) Mathematical symbol for the *turns ratio* of a transformer. (2) Unit symbol for *newton.*

N$_A$ Mathematical symbol for Avogradro's constant, 6.023×10^{28} atoms/mole.

nano Abbreviated *n,* prefix for 10^{-9}.

napier See *neper.*

Napier, John A Scottish mathematician (1550–1617) who invented the logarithm, and after whom the logarithmic unit of measure was named. See *neper, natural logarithm.*

Napierian logarithm See *natural logarithm.*

natural frequency The frequency at which a resonant system oscillates without damping or externally supplied energy.

natural logarithm The base e logarithm. Also called the *Napierian logarithm.*

NC (or nc) (1)Abbreviation for normally closed. (2) Abbreviation for no connection.

near field Any electromagnetic field less than a distance of one wavelength from its source.

neg Abbreviation for *negative.*

negative (1) Abbreviated *neg,* a terminal or electrode that has a negative charge. (2) Opposite of positive.

negative feedback An output-to-input signal path where the output signal is 180 degrees out of phase with the input, leading to a reduction in circuit gain and distortion. Also called *inverse feedback, stabilized feedback,* or *degeneration.*

negative temperature coefficient The amount of reduction in a component's value with a given increase in temperature.

nematic Pertaining to a liquid crystal material in which the charged molecules are parallel to each other, but not in layers.

neon bulb A glass bulb filled with neon gas containing two electrodes. The bulb will not glow until the voltage between the electrodes reaches the firing, or ionization, potential, and remains glowing until the voltage decreases to the extinction level.

neper Abbreviated *Np*, a logarithmic unit of measure used to compare the ratio of two like quantities, such as voltage, current, or power. In terms of the power ratio P_2/P_1,

$$Np = \frac{1}{2} \ln \frac{P_2}{P_1}$$

In terms of the voltage ratio V_2/V_1,

$$Np = \ln \frac{V_2}{V_1}$$

Equal to 8.686 dB. Also called *napier.*

Nernst bridge An AC bridge where all four arms are capacitors, used for measuring capacitances at high frequencies.

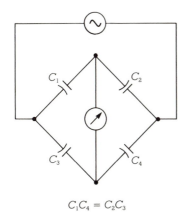

$$C_1C_4 = C_2C_3$$

Nernst bridge

network Any interconnection of electrical or electronic devices.

network synthesis The construction of a network from knowledge of various electrical properties of the network, such as transfer impedance and the output voltage for a specified input voltage.

neutralization The cancellation of voltage feedback from the output to the input of an amplifier through the interelectrode capacitance of a tube or transistor. Neutralization of an amplifier then prevents oscillation.

neutron A part of the nucleus of an atom that has no charge.

newton Abbreviated *N*, the unit of force equal to the force that will cause a mass of 1 kilogram to accelerate at 1 meter per second per second. Equal to 10^5 dynes.

nicad Acronym for nickel-cadmium. Pertaining to a sealed storage battery having a nickel anode, a cadmium cathode, and a potassium hydroxide electrolyte.

nit Abbreviated *nt*, a measure of luminance, equal to 1 candela per square meter.

NO Abbreviation for normally open.

node (1) A point in an electrical network that is common to two or more branches. Also called a nodal point. (2) A point of zero amplitude in a system of standing waves.

noise Any unwanted voltage or current fluctuation in an electrical system.

noise figure Abbreviated *F*, a figure of merit of a system so that:

$$F = \frac{S_{in}/N_{in}}{S_{out}/N_{out}}$$

where *S* and *N* are the power levels of the signal and noise at both the input and output of the system.

noninverting amplifier   An amplifier whose output signal is in phase with its input, such as common-collector, common-base, common-drain, common-gate, common-plate, or common-grid amplifiers.

noninverting input   One of two inputs of an operational amplifier, or comparator.

Norton equivalent circuit   An equivalent circuit between two terminals consisting of an ideal current source in parallel with a Norton (Thevenin) impedance.

Norton's theorem   In a linear network, the current through an impedance connected across a pair of terminals A and B is the same as if the impedance were connected to an ideal current source, equal to the short circuit current between the terminals A and B, and, in parallel with an output impedance, equal to the network impedance between A and B with all voltage sources replaced by their internal impedances.

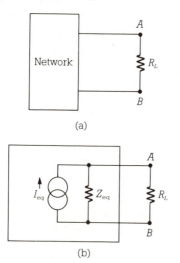

Norton's theorem: (a) General circuit; (b) Norton equivalent circuit

notch filter   A filter network capable of rejecting a single frequency or frequency band within a range of possible input frequencies. Also called a *band-elimination, band-reject,* or *band-stop filter.*

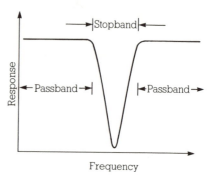

Notch filter response

notch frequency   In a notch filter, the frequency at which the output voltage is at a minimum.

Np   Unit symbol for *neper.*

NPN transistor   A semiconductor device where the collector and emitter are made from n-type semiconductor material while the base is made from p-type material. See *transistor.*

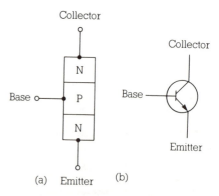

NPN transistor: (a) Internal structure; (b) symbol

nt Unit symbol for *nit.*

***n*th harmonic** A harmonic whose frequency is n times the frequency of the fundamental, where n is an integer greater than 1.

nu (ν) The thirteenth letter of the Greek alphabet, used to designate *frequency.*

null (1) A condition of network balance so that the output is zero. (2) In an antenna radiation pattern, the points where the field strength is zero.

null balance A condition in which two or more signals are combined to give a zero voltage.

null depth In a notch filter, the maximum amount of rejection, usually expressed in decibels. Also called *depth of null.*

octave The interval between two frequencies having a ratio of 2:1.

odd harmonic Any frequency that is an odd multiple of the fundamental frequency.

Oe Unit symbol for *oersted.*

oersted Abbreviated *Oe*, the unit of electromagnetic force in the GCS system. Equal to 79.58 ampere-turns per meter.

Oersted (or Ørsted), Hans Christian A Danish physicist and chemist (1777–1851) who investigated relationships between electricity and magnetism, and after whom the unit of electromagnetic force was named.

offset (1) The component of error that is constant and independent of the inputs. (2) The measure of unbalance between both halves of a symmetrical circuit. (3) The difference between the desired control point and that actually obtained.

offset current In a DC amplifier, such as an operational amplifier, the difference in the current flowing in the two input leads.

offset voltage (1) In a DC amplifier, the output voltage error when there is no input signal. (2) In the absence of any input signal, the DC voltage that must be placed in series with either input of a DC amplifier to produce a zero output voltage.

ohm Symbolized by the Greek letter Ω, the unit of electrical resistance, impedance, and reactance. One ohm is the resistance with which a current of one ampere will maintain a voltage drop of 1 V.

Ohm, Georg Simon A German physicist (1787–1854) who studied electric current, and after whom the unit of electrical resistance was named.

ohmic resistance DC resistance.

ohmmeter An instrument for measuring resistance.

Ohm's Law (1) For electric circuits, the voltage across a resistance is directly proportional to the current flowing through it, or $V = IR$. (2) For magnetic circuits, the magnetomotive force developed across a reluctance is proportional to the magnetic flux, or $F_m = \phi R_m$.

ohms-per-volt In voltage measuring instruments, a measure of meter sensitivity as defined by the meter resistance divided by the full-scale meter voltage.

omega (Ω, or ω) Twenty-fourth and last letter of the Greek alphabet, used to symbolize the *ohm* (Ω), and *angular frequency* (ω).

one-shot See *monostable multivibrator.*

O-network A network having four impedance branches connected in series to form a closed loop in the shape of the letter O whereby two adjacent nodes serve as input terminals while the remaining two nodes serve as output terminals.

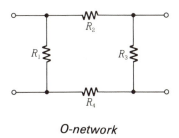

O-network

op-amp Abbreviation, or shortened form for *operational amplifier.*

open (1) A circuit interruption. (2) Infinite resistance.

open circuit (1) A circuit where there is no complete path for current to flow. (2) A circuit or branch having infinite resistance.

open-circuit parameters See *Z-parameters.*

open loop A system or signal path without feedback.

open loop gain The voltage gain of an amplifier connected to a load without any feedback path between its output and its input.

open (frame) motor A motor with ventilation holes for the passage of external cooling air over and around its internal windings.

operating frequency The rated AC frequency of the supply voltage at which a relay is designed to operate.

operating point (1) Quiescent point. (2) For a bipolar transistor, the corresponding DC collector current and collector-emitter voltage. (3) For a field effect transistor, the corresponding DC drain current and drain-source voltage.

operational amplifier Abbreviated *op-amp,* a high-gain DC amplifier with a high input impedance and low output impedance. It can be used to provide amplification, specific computational functions, or signal generation.

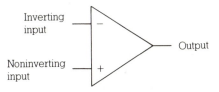

Operational amplifier

OR-gate A digital device whose output is at logic 1 when one, more, or all of its inputs is at logic 1; otherwise its output is at logic 0.

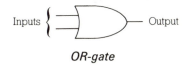

OR-gate

oscillator (1) A self-excited active circuit whose output voltage is a periodic function of time. (2) An amplifier with positive feedback.

oscillograph A recorder, generally using light-sensitive paper, that produces a plot of a time-varying waveform.

oscilloscope An instrument using a cathode-ray tube to display a signal's variation with time. Also called a *scope.*

output (1) The voltage, current, power, or other driving force delivered by a circuit or device. (2) The terminals where a circuit or device may deliver a voltage, current, power, or driving force.

output impedance The impedance measured at the output terminals of a circuit or device with the load disconnected.

output offset voltage The voltage offset at the output of a DC amplifier when the input signal is zero.

output power The power delivered by a system to a load.

output pulse width See *time delay.*

output stage The final stage in any electronic device.

output transformer A transformer used for coupling either the plate or collector circuit of an amplifier to a speaker or other load.

overcompounding Use of sufficient series turns in a compound-wound generator to raise the voltage as the load increases. Consequently, the generator's speed increases as the load increases.

overcoupling Coupling greater than critical.

overdamped (1) The absence of overshoot or oscillation after the application of a sharply rising input signal. (2) Damping greater than the amount required for critical damping.

overdriven amplifier An amplifier, or amplifier stage, whose input signal is large enough to drive it either into saturation or cutoff, thus distorting the input signal.

overload A load greater than a component or system is intended to handle.

overload protection A device which automatically disconnects, or shuts down, a circuit whenever the current or voltage exceeds a safe value.

overmodulation (1) The degree of amplitude modulation where the amplitude of the modulating wave exceeds one-half the amplitude of the unmodulated carrier. (2) Modulation greater than 100 percent.

overshoot A transient behavior which exceeds the steady-state response. Also called *transient overshoot.*

overtone A tone in a complex sound wave that has a frequency higher than the fundamental. The nth overtone is equal to the $n - 1$th harmonic.

Owen bridge A four-arm AC bridge for measuring self-inductance in terms of capacitance and resistance where the balance is frequently independent.

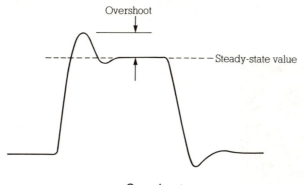

Overshoot

Overshoot

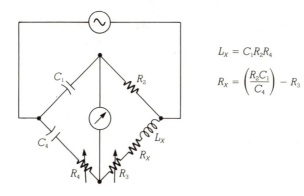

$$L_x = C_1 R_2 R_4$$

$$R_x = \left(\frac{R_2 C_1}{C_4}\right) - R_3$$

Owen bridge

p Prefix symbol for *pico* (10^{-12}).

P (1) Mathematical symbol for *power, permeance.* (2) Prefix symbol for *peta* (10^{15}). (3) Abbreviation for *plate.* (4) Letter designation for a *plug,* or *male connector,* on schematic diagrams.

pA Abbreviation for *picoampere.*

PA (or pa) Abbreviation for *power amplifier.*

pad (1) An arrangement of fixed resistors used to reduce the signal level by a given amount without distortion. (2) A fixed attenuator.

PAM Abbreviation for *pulse-amplitude modulation.*

panel A metallic or nonmetallic sheet on which the operating controls of an electronic device or system are mounted.

parallel circuit A circuit in which the same voltage is applied to all components.

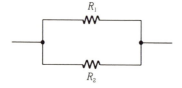

Parallel circuit

parallel feed See *shunt feed.*

parallelogram distortion In a camera or image tube, a form of distortion which amounts to a skewing of the reproduced image laterally across the display.

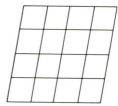

Parallelogram distortion

parallel-resonant circuit A resonant circuit in which the capacitor and inductor are in parallel with the applied AC voltage. At resonance, the reactance of the capacitor and inductor are equal so that the equivalent impedance of the parallel combination is at a maximum. The voltage across the parallel combination will also be at a maximum. Also called an *antiresonant circuit.*

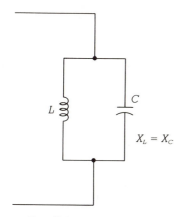

Parallel-resonant circuit

parallel-series circuit Two or more parallel circuits connected together in series. Also called shunt-series circuit.

parallel-T network See *twin-T network*.

paraphase amplifier An amplifier having two output signals which are 180 degrees out of phase with each other, often used to drive a push-pull amplifier stage.

parasitic An unwanted signal or oscillation.

parasitic element An antenna element that serves as part of a directional antenna system, but which has no direct connection to the receiver or transmitter. It reflects or reradiates electromagnetic energy that reaches it.

passband The frequency range of a filter network where the attenuation is at a minimum. Also called *bandpass,* or *filter passband.*

passive Involving energy that is reflected or radiated naturally by an object.

passive device A device that exhibits no gain and does not require any input other than a signal to perform its function. Resistors, capacitors, and inductors are passive devices.

passive network A network with no source of energy.

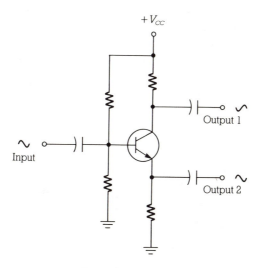

Paraphase amplifier

109

patch To temporarily connect circuits together with a special cord known as a patch cord.

P-band The radio-frequency band from 225 to 390 MHz.

PC (or pc) Abbreviation for *printed circuit.*

PCB Abbreviation for *printed circuit board.*

PCM (or pcm) Abbreviation for *pulse-code modulation.*

peak (1) The maximum instantaneous value of a quantity. Also called *crest.* (2) To broaden the frequency response of an amplifier.

peak amplitude The maximum amplitude of a time-varying quantity as measured from its zero value.

peak clipper A limiter.

peak envelope power Abbreviated *PEP*, the average RF power supplied by a transmitter during one RF cycle at the highest peak-to-peak value of the modulation envelope. It is equal to the input power when the transmitter's power amplifier is driven by a continuous RF signal that has a peak amplitude which the amplifier can handle without distortion.

peak factor For a periodic waveform, equal to the ratio of the peak value to the root mean square value. For a sinewave, it is equal to 1.414. Also called *crest factor.*

peaking network A circuit that improves the high-frequency response of a broadband amplifier. It consists of an inductor placed either in series (series peaking) or in parallel (shunt peaking) with a parasitic capacitance.

peak inverse voltage (1) Abbreviated *PIV*, the peak AC voltage that a semiconductor diode will withstand in the reverse direction. Also called *inverse peak voltage.* (2) Breakdown voltage.

peak-to-peak amplitude The amplitude of a periodic waveform as measured from its positive peak to its negative peak.

peak value The maximum amplitude of a time varying quantity as measured from its zero value. Also called *crest value.*

pentode A vacuum tube that has five elements: cathode, plate, control grid, screen grid, and suppressor grid.

PEP Abbreviation for *peak envelope power.*

percent modulation See *modulation coefficient.*

period Symbolized by T. (1) The time required for one complete cycle of a periodic waveform. (2) The time interval between any two successive similar phases of a periodic waveform.

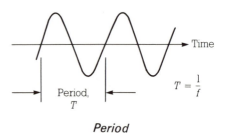

Period

periodic Repeating itself in time and in shape.

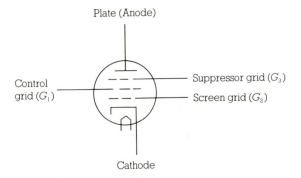

Plate (Anode)

Control
grid (G_1)

Suppressor grid (G_3)

Screen grid (G_2)

Cathode

Pentode

permanent magnet Abbreviated *PM*, a piece of hardened steel or other magnetic material which has been strongly magnetized and retains its magnetism indefinitely.

permeability Symbolized by the Greek letter μ, a parameter of a material which is equal to the magnitude of the magnetic induction (B) divided by the magnetic field strength (H).

permeance Symbolized by E and P_m, and expressed in henries, the reciprocal of reluctance.

permittivity See *dielectric constant.*

peta Abbreviated P, prefix for 10^{15}.

pF Abbreviation for picofarad.

PF (or pf) Abbreviation for *power factor.*

phase (1) The position of a point on a periodic waveform with respect to the start of the cycle and expressed in electrical degrees or radians. (2) The angular relationship between current and voltage in AC circuits. (3) Any one of the single circuits of a polyphase AC power system.

phase angle The angle, expressed either in electrical degrees or radians, that indicates the amount by which the current in an electric circuit leads or lags the voltage. Also called *impedance angle.*

phase compensation network A network used to provide closed-loop stability in an operational amplifier so that the rolloff of the open-loop gain is − 12 dB per octave or less.

phase detector A circuit that provides a DC output voltage that is proportional to the phase difference of its two input signals. Also called a *phase discriminator.*

phase difference The time, expressed in electrical degrees or radians, by which one waveform leads or lags another.

phase discriminator See *phase detector.*

phase inversion The condition whereby the output signal of a circuit, although having the same shape and frequency, is 180 degrees out of phase with the input.

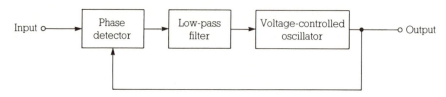

Phase-locked loop

phase lock The condition whereby the phase of an oscillator's output signal exactly follows the phase of a reference signal by comparing the phase difference of the two signals, which in turn controls the frequency of the reference oscillator.

phase-locked loop Abbreviated *PLL*, a feedback control circuit that consists of a phase detector, a low-pass loop filter, and a voltage-controlled oscillator which compares the frequency of the voltage-controlled oscillator with that of an input signal reference. The DC output of the phase detector, after passing through the loop filter, drives the voltage-controlled oscillator to keep it exactly in phase with the input signal reference.

phase margin The additional amount of phase shift of the output signals of a feedback control loop at the open-loop, unity-gain crossover frequency that would produce instability.

phase modulation Abbreviated *PM*, modulation in which the angle of the sinewave carrier deviates from the original angle by an amount proportional to the modulating signal's amplitude.

phase quadrature See *quadrature*.

phase reversal A change of one-half cycle, or 180 degrees, in phase.

phase shift (1) A change in the phase of a periodic quantity. (2) The difference between corresponding points on input and output signals.

phase shift oscillator An oscillator in which a network that has a phase shift of 180 degrees is connected between the output and input of an amplifier.

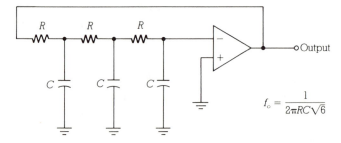

$$f_o = \frac{1}{2\pi RC\sqrt{6}}$$

Phase shift oscillator

phasor A quantity expressed in complex form, with or without time variation. A phasor may be used to represent a vector, but a vector does not involve the complex plane and therefore is not a phasor.

phi (Φ, or φ) The twenty-first letter of the Greek alphabet, used as a symbol for *angles* (φ), *phase angle* (φ), or *magnetic flux* (Φ).

phone plug A plug used with headphones, microphones, and other audio equipment. Also called a *telephone plug*.

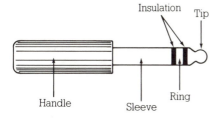

Phone plug

phono plug A type of plug used for AF and low-level RF shielded signal cables.

Phono plug

photocell A solid-state photosensitive device whose transfer characteristic is a function of the incident light intensity. Also called a *photoelectric cell,* or *electric eye.*

photoconductive cell A photocell whose internal resistance varies with the incident light intensity on the cell's active material.

photoelectric cell See *photocell.*

photon A quantum of electromagnetic radiation, equal to the product of Planck's constant and the frequency in hertz.

pi (π) The sixteenth letter of the Greek alphabet, used to designate the value of the irrational number 3.14159, which is equal to the ratio of the circumference of a circle to its diameter.

pico Abbreviated p, prefix for 10^{-12}.

Pierce oscillator A modification of a Colpitts oscillator in which a piezoelectric crystal replaces the inductor in the tank circuit between the collector (plate) and base (grid) of the transistor (vacuum tube) circuit.

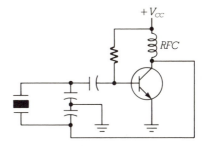

Pierce oscillator

piezoelectric effect The production of a voltage between opposite faces of a piezoelectric crystal as a result of strain due to pressure. The reverse effect also occurs in that the application of a voltage to the opposite faces of a piezoelectric crystal causes deformation or vibration at the frequency of the applied voltage.

pigtail A short, flexible wire, either stranded or braided, used between a stationary terminal and a terminal that has a limited range of motion, such as a relay armature.

pinch-off voltage The voltage at which the current flow between the source and drain in a field effect transistor is blocked due to the depleted channel between these electrodes. For n-channel field effect transistors the pinchoff voltage is positive; for p-channel types it is negative.

pi-network See *delta network.*

PINO Acronym for positive input, negative output.

PIV Abbreviation for *peak inverse voltage.*

Planck's constant Symbolized by the letter *h*, a physical constant equal to 6.6256×10^{-34} joule-second.

plate (1) Symbolized by the letters *P* or *A*, the positive terminal of an electron tube, also called the *anode.* (2) One of the conducting surfaces in a capacitor.

PLL Abbreviation for *phase-locked loop.*

plug Designated by the letter *P* on schematic diagrams, a device on the end of a cable that will fit into a jack. The plug is generally thought of as the male portion of the plug-jack connection.

PM (or pm) Abbreviation for *permanent magnet* or *phase modulation.*

PN junction A two-terminal, solid-state junction made from a semiconductor that has been treated to conduct current more readily in one direction than in the other. This treatment results in a p-type semiconductor at one end and an n-type semiconductor at the other end. The PN junction forms the basis of most solid-state devices, such as diodes, transistors, and silicon controlled rectifiers.

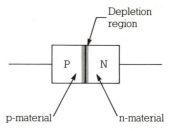

P-N junction

PNP transistor A semiconductor device where the collector and emitter are made from p-type semiconductor material while the base is made from n-type material. In normal operation the emitter is positive with respect to the base while the collector is negative with respect to the base. See *transistor.*

polarized plug A multiple-pin plug that can be inserted into its socket or receptacle only in predetermined orientation.

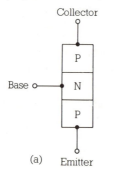

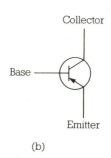

(a)　　　　　　　　　(b)

PNP transistor: (a) Internal structure; (b) Symbol

pole (1) A region in a magnet that has polarity, such as the north or south pole. (2) An output terminal on a switch. (3) A point at which the denominator of a function becomes zero so that the function approaches infinity.

pholyphase motor An induction motor wound for operation on two- or three-phase AC voltage.

popcorn noise Noise produced by erratic jumps of bias current at random intervals in operational amplifiers and other semiconductor devices.

pos Abbreviation for *positive.*

positive feedback An output-to-input signal path where the output signal is fed back in phase with the input in order to increase amplification. Excessive positive feedback causes distortion, instability, and, when sufficiently high, oscillation. Also called *regeneration,* or *regenerative feedback.*

pot Slang term for *potentiometer.*

potential (1) Symbolized by *V*, the degree of electrification as referenced to some standard such as ground. Also referred to as *voltage.* (2) At a given point in space, equal to the work required to bring a unit quantity of electricity from infinity to that point.

potential difference The voltage that exists between two given points in a circuit, such as across a circuit element.

potentiometer (1) A three-terminal resistor that has an adjustable sliding contact or wiper so that its resistance may be varied. Also called a *pot.* (2) A variable resistor. (3) Any instrument in which a potentiometer is

used as a voltage divider in order to produce a known voltage that can be balanced against an unknown voltage.

Potentiometer

power (1) Symbolized by the letters *P* or *E*, the time rate of change of work. (2) Measured in watts, the product of voltage and current. (3) The value of an exponent.

power amplifier Abbreviated *PA*, an AF or RF amplifier that delivers maximum output power to a load rather than providing a maximum voltage gain.

power cord See *line cord.*

power dissipation The release of the heat generated within a device when a current flows through it.

power factor (1) Abbreviated *PF*, the ratio of active power to apparent power. (2) The cosine of the phase angle between the voltage and current in an AC circuit load.

power gain Abbreviated A_p, the ratio of the output power of a network to the applied power. Expressed in decibels, the power gain is

$$A_p = 10 \log \frac{P_2}{P_1}$$

power line Two or more wires conducting electric power from one location to another. Called *mains* in Britain.

power output The power in watts delivered to a load.

power rating The maximum power that can be dissipated in a component or device that is operated according to manufacturer's specifications for a specified period.

power ratio See *power gain.*

power supply A device that supplies electrical power, normally DC power.

power transformer An iron-core transformer in which the primary winding is connected to the AC power line while having one or more secondary windings that provide voltages less than or greater than the power line.

power transistor A transistor capable of handling high currents and power, generally greater than 1 watt.

preamplifier An amplifier intended to operate with low-level signals to provide gain and impedance matching to a level that can be handled by another amplifier. Also called *preamp.*

preemphasis See *emphasis.*

preemphasis network An RC filter used to emphasize one range of frequencies with respect to another.

prefix multipliers Prefixes that designate a greater or smaller unit than the original by the factor indicated in the Supplement, "Multiplier Prefixes," page 180.

pri Abbreviation for *primary.*

primary Abbreviated *pri* or *P,* the input windings of a transformer. Also called primary winding.

printed circuit Abbreviated *PC,* a circuit in which the connecting wires are replaced by conductive strips or foils that have been printed or etched onto an insulated board.

printed circuit board Abbreviated *PCB,* an insulated board, generally made from epoxy or bakelite, on which a circuit has been printed and components mounted.

probe A test prod used with test equipment.

programmable unijunction transistor Abbreviated *PUT,* a three-terminal, solid-state device having an anode, a cathode, and a gate. Its action is similar to that of a silicon controlled switch in that for a constant gate voltage the device remains nonconductive until the anode voltage is greater than the gate voltage by a predetermined amount, after which it conducts.

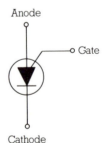

Programmable unijunction transistor

propagation loss The attenuation of signals passing between two points on a transmission line or waveguide.

propagation time delay The time required for a wave to travel between two points on a transmission line or waveguide.

proton An atomic particle located in the nucleus of an atom that has a positive charge equal in magnitude to the negative charge of an electron.

psi (Ψ, or ψ) The twenty-third letter of the Greek alphabet, used as a symbol for *electric flux.*

pulse A momentary, sharp change in a current, voltage, or other quantity that is normally constant, characterized by a rise and fall of finite duration.

pulse-amplitude modulation Abbreviated *PAM*, amplitude modulation of a pulse train used as a carrier signal.

pulse-code modulation Abbreviated *PCM*, modulation in which the

signal is periodically sampled and each sample is then grouped and transmitted as a digital binary code.

pup jack See *tip jack.*

push-pull amplifier A balanced amplifier that uses two similar vacuum tubes or transistors working in phase opposition.

push-push amplifier An amplifier that uses two similar vacuum tubes or transistors with their grid or base leads connected in phase opposition while the plate or collector leads are connected in parallel to a common load. It is used primarily as a frequency multiplier to emphasize even-order harmonics.

PUT Abbreviation for *programmable unijunction transistor.*

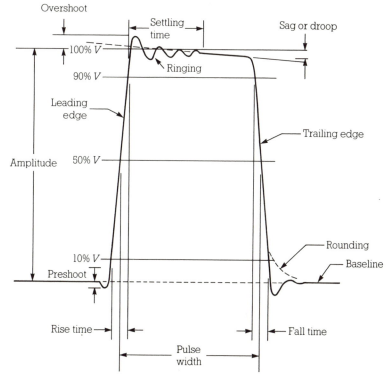

Pulse characteristic parameters

Q (1)Abbreviation for *quality factor*.
(2) Symbol for *electric charge*.
(3) Letter designation for a *transistor*
on a schematic diagram.

Q-band The radio-frequency band
from 36 to 46 GHz.

Q-channel The 0.5 MHz-wide band
used in the American NTSC color television system for transmitting green-
magenta color information.

Q-factor Same as *Q, quality factor*.

Q-meter An instrument for measuring the *Q* of a circuit or circuit element. Also called a quality-factor
meter.

Q-multiplier A filter circuit which
gives either a very sharp peak or
rejection at a particular frequency,
which is equivalent to increasing the
Q of a tuned circuit at that frequency.

quadrant A 90-degree sector of a
circle.

quadrature The relation between
two periodic signals when the phase
difference between them is 90
degrees or one-quarter cycle. Also
called *phase quadrature*.

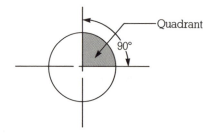

Quadrant

quadrature amplifier An amplifier
that shifts the phase of a signal by 90
degrees.

quadrature modulation The modulation of two carrier signals 90
degrees apart by separate modulating functions.

quality factor Abbreviated *Q* or *Q-
factor*, a dimensionless figure of merit
or selectivity for a resonant circuit or
system. The *Q* of an inductor at a
given frequency is the ratio of its
reactance to its series resistance, or
$2\pi fL/R$. The *Q* of a capacitor at a
given frequency is its susceptance
times its effective shunt resistance.
The *Q* of a simple LC resonant circuit
is $Q_L Q_C/(Q_L + Q_C)$. The *Q* of a bandpass, or notch filter, network is the ratio of its resonant, or center, frequency to its bandwidth, f_o/BW.

quarter phase See *two-phase.*

quarter-wave antenna An antenna whose electrical length is one-fourth the wavelength of the transmitted or received signal.

quarter-wavelength The distance that corresponds to an electrical length of one-fourth the wavelength at the operating frequency of a transmission line or antenna element.

quarter-wave line See *quarter-wave stub.*

quarter-wave resonance In a quarter-wave antenna, the condition in which its resonant frequency equals the frequency at which it is used.

quarter-wave stub A section of transmission line equal to one-fourth the wavelength at the fundamental frequency. The far end is shorted so that the open end appears as a high impedance at the fundamental frequency and all odd-order harmonics, but not at even-order harmonics. Also called a *quarter-wave line* or *quarter-wave transmission line.*

quarter-wave transformer A quarter-wave section of transmission line used for impedance matching.

quarter-wave transmission line See *quarter-wave stub.*

quartz A mineral form of silicon dioxide that forms hexagonal crystals having piezoelectric properties. When excited electrically, they vibrate and maintain extremely accurate and stable frequencies.

quartz crystal A thin slab cut from quartz and ground to the thickness at which it will vibrate at a given frequency when electrically excited. It is used to accurately control the frequency of an oscillator.

quiescent (1) At rest. (2) The condition of a circuit when no input signal is applied.

quiescent operating point The resultant corresponding voltage and current on the characteristic curve of a vacuum tube or semiconductor device when there is no input signal.

quieting The degree in noise voltage at the output of an FM receiver in the presence of an unmodulated carrier.

r   Mathematical symbol for *AC resistance*.

R   (1) Mathematical symbol for *DC resistance*. (2) Letter designation for a *resistor* on schematic diagrams.

raceway   A protective channel used to hold busbars, cables, or wires.

rad   Abbreviation and unit symbol for *radian*.

radar   Acronym for radio detecting and ranging, a system that uses beamed and reflected RF energy for the detection and location of objects, as well as for measuring distances or altitudes.

radian   Abbreviated *rad*, the angle that intercepts an arc whose length is equal to its radius, so that a complete circle contains 2π radians. Equal to 57.29579 degrees.

radian frequency   Expressed in radians per second, equal to 2π times the frequency in hertz.

radiant energy   Energy transmitted in the form of electromagnetic radiation, such as heat, light, or radio waves.

radiation angle   The angle between the surface of the earth and the center of the beam of energy radiated upward into the sky from a transmitting antenna. Also called *angle of radiation*.

radiation pattern   The graphical representation of the radiation of an antenna as a function of direction. Also called *field pattern*.

radiation resistance   The ratio of the total power radiated by an antenna divided by the square of the effective antenna current as measured at the point where power is supplied to the antenna.

radiator   (1) In an antenna, the part that radiates electromagnetic waves either directly into space or against a reflector for its direction. (2) A body or surface that emits radiant energy.

radio   (1) A term used to describe the transmission of electromagnetic waves through space. (2) A receiver apparatus. (3) A prefix that denotes the use of radiant energy in the form of radio waves.

RF Region	Frequency Range	Wavelength
VLF	10–30 kHz	10–100 km
LF	30–300 kHz	1–10 km
MF	300 kHz–3 MHz	100–1 000 m
HF	3–30 MHz	10–100 m
VHF	30–300 MHz	1–10 m
UHF	300 MHz–3 GHz	10–100 cm
SHF	3–30 GHz	1–10 cm
EHF	30–300 GHz	1–10 mm

radio frequency Abbreviated *RF*, electromagnetic radiation having any frequency in the region from 10 kHz to 300 GHz. Also called *radio spectrum.* This range is divided into eight regions as in the table above.

radio-frequency (or RF) amplifier An amplifier that amplifies radio frequency signals. In a receiver, it is sometimes referred to as the *front end.*

radio-frequency (or RF) choke Abbreviated *RFC*, an inductor used to specifically block the flow of RF current.

radio-frequency (or RF) interference Abbreviated *RFI*, interference from sources of radio-frequency energy located outside a system.

radio-frequency (or RF) oscillator An oscillator that generates AC signals at radio frequencies.

radio spectrum The entire range of frequencies, from approximately 10 kHz to 300 GHz, in which radio waves can be produced. Also called *radio frequencies.*

radio tube See *electron tube.*

radio wave A radio-frequency electromagnetic wave traveling at the speed of light. Also called *Hertzian wave.*

rain voltage British term for *supply voltage.*

RAM Acronym for *random access memory.*

ramp A type of waveform that changes value at a constant rate, such as a sawtooth waveform.

random access memory Abbreviated *RAM*, memory that information can be either stored in or transferred from with a speed that is independent of the location of the information in storage.

RC (or R-C) Abbreviation for resistor-capacitor, resistance coupled, or remote control.

RC constant See *RC time constant.*

RC filter Any passive filter network made up of only resistors and capacitors.

RC network Any circuit made up of resistors and capacitors arranged to perform a specific function.

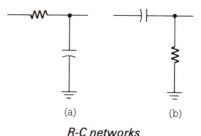

(a)   (b)

R-C networks

121

RC time constant Symbolized by the Greek letter τ, the time required in an RC circuit for a circuit parameter, such as voltage or current, to rise to $1 - 1/e$ (approximately 63 percent), or decay to $1/e$ (approximately 37 percent) of its steady-state value after a step function change in excitation. It is equal to the product of the resistance in ohms and the capacitance in farads. Also called *RC constant.*

reactance (1) Symbolized by the letter X and expressed in ohms, the imaginary part of impedance, which is dependent on frequency. For an inductor, the reactance, X_L, is equal to $2\pi fL$. For a capacitor, the reactance, X_C, is equal to $1/2\pi fc$.
(2) Impedance not due to resistance.

reactance factor The ratio of the AC resistance to the DC, or ohmic, resistance of a conductor.

reactive factor The ratio of reactive power to apparent power.

reactive load As opposed to a resistive load, a load that has inductive or capacitive reactance.

reactive power Also called wattless power and measured in vars, the power in an AC circuit equal to

$$P_{reactive} = V_{RMS} I_{RMS} \sin \theta$$

reactive volt-ampere meter See *varmeter.*

reactor A device that introduces either capacitive or inductive reactance into a circuit.

real number A number whose square is positive.

real power The power in an AC circuit, expressed in watts, equal to the apparent power times the power factor, or

$$P_{real} = V_{max} I_{max} \cos \theta$$

receiver A device that converts electromagnetic waves into a visible or audible form. Also called a radio receiver.

recharger A DC power supply that recharges nickel-cadmium or other rechargeable batteries or cells.

reciprocal energy theorem A theorem proposed by Rayleigh which states that if a voltage V_1 in one branch of a circuit produces a current I_2 in any other branch, and if a voltage V_2 inserted in this other branch produces a current I_1 in the first branch, then $V_1 I_2 = V_2 I_1$ and is related to the reciprocity theorem.

reciprocal impedance Two impedances Z_1 and Z_2 are said to be reciprocal impedances with respect to another impedance Z if they satisfy the relation $Z_1 Z_2 = Z^2$.

reciprocity A property of a passive, linear, bilateral network whereby the interchange of its output and input does not affect the system response for a given excitation. For such a four-port network, the following relationship holds:

$$\frac{V_1}{i_2} = Z_{12} = Z_{21} = \frac{V_2}{i_1}$$

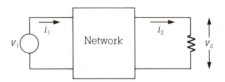

Reciprocity

reciprocity theorem If a network is composed solely of passive, linear, bilateral impedances, the ratio of the voltage introduced into one branch to the current produced in another branch is unchanged if the position of the current and voltage are interchanged.

rectification The process of converting AC to DC.

rectifier (1) A device that converts AC into DC. (2) A device which has the ability to pass current in only one direction.

rectilinear Following a straight line.

reed relay A relay that has contacts mounted on magnetic reeds sealed inside a small glass tube. An actuating coil is wound either around the tube or on a ferrite core to provide the magnetic field to open and close the reeds. The contacts may be either dry or mercury-wetted. Also called a *magnetic reed relay*.

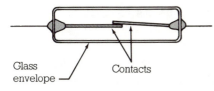

Glass envelope / Contacts

Reed relay

reed switch The contact assembly of a reed relay.

reference dipole A straight, half-wave, dipole antenna designed to operate at a specific frequency and used as a comparison for antenna measurements.

reference level The level used as a basis for comparison of signals in terms of decibels or volume units.

reflected impedance The impedance seen at the input of a network when its output is terminated in an impedance of a specified value. In a transformer, the reflected impedance depends on the load resistance and the transformer's turns ratio, so that

$$Z_{reflected} = \left(\frac{N_1}{N_2}\right)^2 Z_L$$

$$Z_{Reflected} = \left(\frac{N_1}{N_2}\right)^2 Z_L$$

Reflected impedance

reflected wave A wave reflected from a surface, discontinuity, or junction of two different media, such as the wave that travels back to the source end of a mismatched transmission line.

reflection coefficient Symbolized by the letter k, the ratio of the amplitude of the reflected wave along a transmission line to that of the incident wave at a discontinuity in the line or medium. For impedances Z_1 and Z_2 on both sides of the discontinuity, the reflection coefficient is computed from

$$k = \frac{Z_1 - Z_2}{Z_1 + Z_2}$$

Also called *mismatch factor, reflection factor,* or *transition factor.*

reflection factor See *reflection coefficient.*

reflection law The incident angle equals the reflection angle.

reflector A single rod, system of rods, metal screen, or sheet used behind an antenna to increase its directivity.

reflector element A single rod or other parasitic element serving as a reflector in an antenna array.

regeneration *Positive feedback,* also called *regenerative feedback.*

regenerative detector A vacuum tube (or transistor) detector circuit in a radio receiver in which RF energy is fed back from the plate (collector) circuit to the grid (base) circuit to give positive feedback at the carrier frequency, which thereby increases amplification and sensitivity of the detector.

regenerative feedback See *positive feedback.*

regenerative receiver A radio receiver that uses a regenerative detector.

regulator A device for holding constant the voltage or current produced by a generator, alternator, or other source. See also *voltage regulator.*

rejection band See *stop band.*

rejector See *trap.*

rel A unit of reluctance equal to 1 ampere-turn per maxwell.

relaxation inverter An inverter that uses a relaxation oscillator circuit to convert DC power to AC power.

relaxation oscillator An oscillator circuit whose fundamental output frequency is determined by the time of charging or discharging a capacitor or coil through a resistor, thereby producing rectangular or sawtooth waveforms.

relay Symbolized by the letter K on schematic diagrams, an electromagnetically operated mechanical switch.

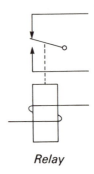

Relay

relay armature The movable iron part of a relay.

relay coil One or more windings on a common form, used with an iron core to form a relay magnet.

relay contacts Contacts that are either closed or opened by the movement of a relay armature.

relay magnet The electromagnet that attracts the armature of a relay when energized.

release time The total elapsed time from the instant the current in the relay coil starts to drop until either the make contacts have opened or the break contacts have closed.

reluctance Symbolized by R_m, expressed in units of reciprocal henry, equal to the ratio of the magnetomotive force in a magnetic circuit to the magnetic flux through any cross-section of the magnetic circuit.

reluctance motor   A synchronous motor, similar in construction to an induction motor, in which the element carrying the secondary circuit has salient poles but no DC excitation. It starts as an induction motor but operates normally at synchronous speed.

remote-cutoff tube   An electron tube, used principally in RF amplifiers, whose control grid wires are further apart at the center than at the ends. Consequently, the tube's amplification is not directly proportional to the bias. Also called a *variable-mu* or *extended-cutoff tube.*

repulsion-induction motor   A repulsion motor that has a squirrel-cage winding in the rotor in addition to a repulsion-motor winding.

repulsion motor   An AC motor that has stator windings connected directly to the AC power source, and rotor windings connected to a commutator. Brushes on the commutator are short-circuited and are positioned to produce the rotating magnetic field required for starting and running. The motor speed varies considerably as the load is changed.

residual field   The magnetic field left in an iron core after excitation has been removed.

residual flux density   The magnetic induction that is left in magnetic material after a saturating magnetizing force is removed. Also called *residual induction, residual magnetic induction,* or *residual magnetism.*

residual induction   See *residual flux density.*

residual magnetic induction   See *residual flux density.*

residual magnetism   See *residual flux density.*

residual modulation   See *carrier noise level.*

residual resistance   The portion of the electrical resistance of a metal that is independent of temperature.

resistance   Abbreviated and symbolized by the letter R, the property of conductors that, depending on their dimensions, material, and temperature, determines the current produced by a given potential difference. Measured in ohms, resistance is defined as the opposition to current flow through which a difference of one volt results from a current of one ampere.

resistance bridge   See *Wheatstone bridge.*

resistance drop   The voltage drop between two points on a conductor when current flows through the resistance between these two points. Using Ohm's law, the voltage drop is equal to the current times the resistance.

resistance loss   The power lost when current flows through a resistance.

resistance noise   See *Johnson noise.*

resistance pad   A network made only of resistors, used to match impedances with a fixed attenuation without changing the frequency response.

resistance ratio   The ratio of the resistances of a thermistor at two specified temperatures, but with no power dissipated by the thermistor.

resistance wire Wire made from a metal or alloy, such as Nichrome, which has a high resistance per unit length, used in wire-wound resistors and heating elements.

resistive load A load in which the current and the voltage are in phase.

resistivity (1) Symbolized by the Greek letter ρ, the resistance in ohms that a unit volume of a material offers to the flow of current, measured in ohm-centimeters. (2) Equal to the reciprocal of conductivity and equal to the resistance multiplied by the cross-sectional area of the material and divided by the length. Also called *specific resistance.*

resistor Symbolized by the letter *R* on schematic diagrams, a device that has a definite amount of resistance, used in circuits to limit current flow or to provide a voltage drop.

resistor color code A method of marking the value in ohms on a resistor by bands of colors as specified by the EIA color code.

resistor core The insulating support on which a resistor element is wound or placed.

resolution A measure of ability to delineate detail, or to distinguish between nearly equal values of a quantity. Also called *resolving power.*

resolver A synchro or other device whose rotor is mechanically driven to translate the rotor angle into a voltage corresponding to the sine and cosine of the rotor angle. Also called *sine-cosine generator* and *synchro resolver.*

resolving power See *resolution.*

resonance The condition of a circuit when the inductive reactance equals the capacitive reactance.

resonance bridge A four-arm AC bridge that measures inductance, capacitance, or frequency. The inductor and capacitor, which may be either in series or parallel, are tuned to resonate at the frequency of the source before the bridge is balanced.

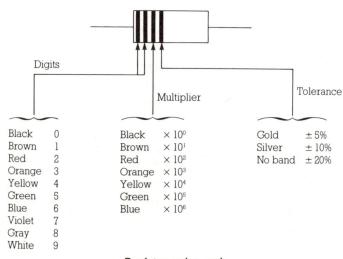

Digits		Multiplier		Tolerance	
Black	0	Black	$\times 10^0$	Gold	± 5%
Brown	1	Brown	$\times 10^1$	Silver	± 10%
Red	2	Red	$\times 10^2$	No band	± 20%
Orange	3	Orange	$\times 10^3$		
Yellow	4	Yellow	$\times 10^4$		
Green	5	Green	$\times 10^5$		
Blue	6	Blue	$\times 10^6$		
Violet	7				
Gray	8				
White	9				

Resistor color code

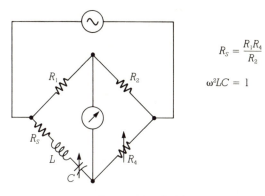

$$R_S = \frac{R_1 R_4}{R_2}$$

$$\omega^2 LC = 1$$

Resonance bridge

resonant circuit   A circuit exhibiting resonance.

resonant frequency   (1) The frequency at which the inductive reactance of a given resonant circuit equals the capacitive reactance. (2) The frequency at which a quartz crystal, loudspeaker diaphragm, or other object will vibrate readily. For an LC circuit the resonant frequency in hertz is determined from

$$f_o = \frac{1}{2\pi\sqrt{LC}}$$

resonant-reed relay   A reed relay in which the reed switch closes only when the required frequency is applied to the operating coil so that one of the reeds vibrates with sufficient amplitude to make contact with the other reed.

response curve   A graph of a characteristic of a circuit, such as gain, as a function of frequency.

restorer   See *DC restorer.*

retrace   See *flyback.*

reverse bias   A bias voltage applied to a semiconductor junction in a direction so that no current flows. Also called *back bias.*

reverse direction   The direction of higher resistance to current flow in a diode or semiconductor junction.

reverse resistance   The resistance measured at a specific reverse voltage drop and current in a diode or semiconductor junction.

reverse voltage   A voltage applied in the reverse direction across a diode or semiconductor junction.

reversible motor   A motor whose rotation can be reversed by a switch that changes the motor connections when the motor is stopped.

reversing motor   A motor whose direction of rotation can be reversed only while the motor is running at full speed. The motor will come to a stop, reverse, and attain full speed in the opposite direction.

RF (or rf)   Abbreviation for *radio frequency.*

RF amplifier   See *radio-frequency amplifier.*

RFC   Abbreviation for *radio-frequency choke.*

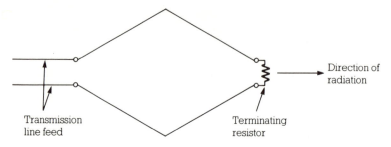

Rhombic antenna

RF choke   See *radio-frequency choke.*

RFI   Abbreviation for *radio frequency interference.*

RF oscillator   See *radio-frequency oscillator.*

RF probe   A detecting device used with a VTVM or other high-impedance voltmeter to measure RF voltages.

rheostat   A variable resistor with one fixed terminal and one terminal connected to a wiper arm.

rho (ρ)   Seventeenth letter of the Greek alphabet, used as a symbol for *resistivity, volume density.*

rhombic antenna   A horizontal, directional antenna that has four sides

forming a diamond or rhombus. It is fed at one apex and terminated with a resistor or impedance at the opposite apex. Also called a *Bruce antenna.*

rig   Slang term for *transmitter.*

right-hand rule   See *Fleming's rule.*

right-hand screw rule   See *Ampere's rule.*

right-hand taper   Taper in which there is greater resistance in the clockwise half of the operating range of a potentiometer or rheostat, looking from the shaft end, than in the counterclockwise half.

ringing   An oscillatory transient that occurs in the output of a system as a result of a sudden change in input.

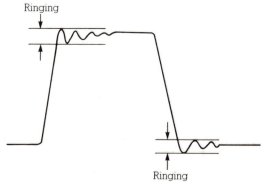

Ringing

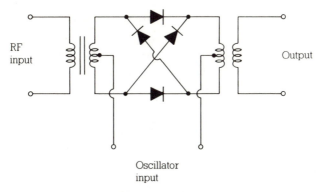

Ring modulator

ring modulator A circuit in which four diodes are connected in series to form a ring around which current flows readily in only one direction. The input and output connections are made to the four nodes of the ring. Used as a balanced modulator, demodulator, or phase detector.

ripple The AC component in the output of a DC power supply as a result of poor filtering, or from commutator action in a DC generator.

ripple factor The ratio of the RMS value of the AC component of a pulsating DC waveform to the average value of the waveform. Usually expressed as a percentage.

ripple frequency The frequency of the ripple present in the output of a DC source.

ripple voltage The AC voltage component of the output of a DC power supply.

rise time Symbolized t_r, the amount of time for a waveform to increase from 10 to 90 percent of its maximum value.

RLC Abbreviation for resistor-inductor-capacitor.

RMA Abbreviation for Radio Manufacturers Association, now called the Electronics Industries Association (EIA).

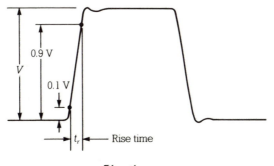

Rise time

RMS (or rms) Abbreviation for *root-mean square.*

rolloff The rate at which the response of a system changes with frequency. Usually expressed in either decibels per decade or decibels per octave.

ROM Acronym for *read-only memory.*

root-mean square Abbreviated *RMS*, the value of an AC waveform that corresponds to the DC value that will give the same heating effect. Also called the *effective value.* Mathematically, it is determined from:

$$V_{\text{RMS}} = \left(\frac{1}{T} \int_0^t v(t)^2 \, dt \right)^{1/2}$$

rotary converter A dynamotor.

rotor (1) The rotating member of a machine or device, such as the rotating armature of a motor or generator. (2) The rotating plates of a variable capacitor.

RTMA Abbreviation for Radio-Television Manufacturers Association, now called the Electronics Industries Association (EIA).

RTTY Abbreviation for radioteletype.

runaway See *thermal runaway.*

s Unit symbol for seconds.

S (1) Unit symbol for *siemen*.
(2) Letter symbol used to designate a *switch* on schematic diagrams.
(3) Abbreviation for *source, secondary*.

salient pole A magnetic field pole, on which is mounted a field coil of a generator or motor.

sample and hold A technique of artificially increasing the duration of a signal pulse, or converting from analog to digital form, by sampling at fixed intervals and storing (holding) amplitudes for display or conversion.

sampling Obtaining values of a function that corresponds to discrete regular or irregularly spaced intervals of the independent variable.

sampling oscilloscope An oscilloscope in which fast and repetitive signals are slowed down for conventional display.

sampling theorem Used in information theory, a method whereby a continuous waveform having a given frequency f can be statistically repro-

duced by sampling it at a rate of every $1/2f$ seconds.

saturation (1) The condition in which a further increase in one quantity produces no further increase in a dependent quantity. (2) In color television, the degree to which color is mixed with white so that the more white, the lower the saturation.

saturation curve A curve that shows the manner in which a quantity, such as current or magnetic flux, reaches saturation as a function of another quantity.

sawtooth A waveform characterized by a slow, linear rise time and virtually an instantaneous fall time, resembling the teeth of a saw. *(See figure on page 132.)*

S-band The radio-frequency band from 1.55 to 5.2 GHz.

SCA Abbreviation for *Subsidiary Communications Authorization*.

scalar Having magnitude but not direction.

scalar product See *dot product*.

schematic

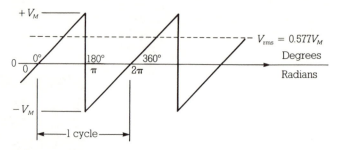

Sawtooth waveform

schematic A *circuit diagram* in which component parts are represented by simple, easily drawn symbols. Also called a schematic circuit diagram.

Schering bridge A four-arm AC bridge that is used to measure capacitance and dissipation factor, with the bridge balance being independent of frequency.

Schmitt trigger A bistable multivibrator that converts an AC input signal into a square wave output by switching action, triggered at predetermined setpoints on the positive and negative swings of the input signal.

Schottky barrier See *Schottky diode.*

Schottky diode A semiconductor diode formed by contact between a semiconductor layer and a metal coating, which has a nonlinear rectification characteristic. Also called a *Schottky barrier,* or *hot-carrier diode.*

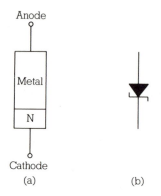

Schottky (hot carrier) diode: (a) Internal structure; (b) Symbol

$$C_X = \frac{C_4 R_2}{R_1}$$

$$D_X = \omega C_X R_X = \omega C_2 R_2$$

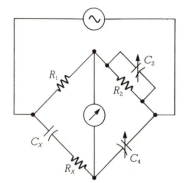

Schering bridge

scope   Slang term for *oscilloscope* or any cathode-ray tube instrument.

SCR   Abbreviation for *silicon controlled rectifier.*

screen   (1) The surface of a cathode-ray tube on which a visible image appears. (2) A metal partition used to shield circuits or components from electric or magnetic fields. (3) A screen grid in a vacuum tube.

screen grid   Abbreviated *SG*, a vacuum tube element positioned between the control grid and the plate. It is kept at a fixed potential in order to reduce the electrostatic effect of the plate element. Also called the *second grid* or G_2.

SCS   Abbreviation for *silicon controlled switch.*

secondary   Abbreviated *S*, the output winding of a transformer.

second detector   The demodulator stage in a superheterodyne receiver.

second grid   See *screen grid.*

section   A filter section.

Seebeck effect   The effect whereby a voltage is produced at a junction of two dissimilar metals when heated; the basis of the thermocouple. Also called *thermoelectric effect.*

selectivity   The characteristic of a circuit or radio receiver that distinguishes between the desired frequency and other nearby frequencies.

selenium rectifier   A rectifier diode made from a junction of a selenium anode and a cadmium cathode.

semiconductor   A crystalline material, such as germanium or silicon, whose electrical conductivity is between that of a conductor and an insulator.

sensitivity   A figure of merit that expresses the ability of a circuit or device to respond to an input quantity.

sensor   A generalized term for any device that detects a change in a physical quantity, such as light or vibration, and converts this change into a signal useful for amplification, control, or measurement.

series   A connection of two or more circuit elements end to end so that the same current flows through each element.

series feed   The application of a DC voltage to the plate of a vacuum tube, collector of a transistor, or drain of a field effect transistor through the load that is carrying the output signal.

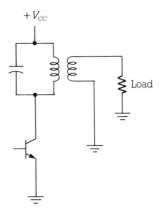

Series fed circuit

series motor   A motor in which the field and armature circuits are connected in series. Also called a *universal motor.*

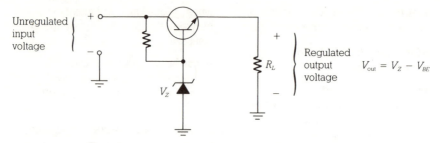

Series regulator

series regulator A device, such as a transistor, placed in series with a power source that is able to automatically vary its series resistance, which in turn controls the output voltage or current.

series resonant circuit A circuit in which an inductor and capacitor are connected in series and have values such that the reactance of the inductor equals the reactance of the capacitor at the resonant frequency, characterized by maximum current flow through the capacitor and inductor.

series-shunt network See *ladder network*.

series-wound motor A commutator motor in which the field and armature circuits are in series.

serrodyne A frequency converter or mixer in which the output of the local oscillator is a sawtooth waveform, which in turn is used to phase modulate the input.

settling time The time required for an error signal in a feedback control system to decrease to a given percentage of its peak value. Typically this percentage can range from one to five percent.

seven-segment display A liquid crystal or light-emitting diode display composed of seven individual segments arranged in figure-eight pattern to form any number from 0 to 9 when the individual segments are lit. The individual segments are lettered *a* through *g* in a clockwise fashion.

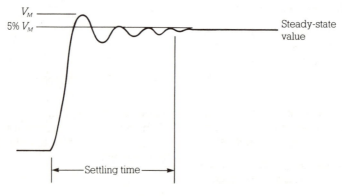

Settling time

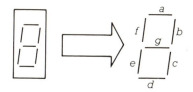

Seven-segment display

SG (or sg) Abbreviation for *screen grid.*

shape factor A dimensionless quantity used to quantify the selectivity of a broadband filter or tuned amplifier stage. It is equal to the ratio of the 60-dB bandwidth to the 6-dB bandwidth.

shell-type transformer A power transformer in which the primary and secondary windings are wound over each other on the center section of an iron core.

SHF (or shf) Abbreviation for *super high frequency.*

shield A metallic partition, either solid, braided, or meshed, used to protect circuits, transmission lines, and cables from stray voltages or currents induced by electric or magnetic fields.

shielded cable A single or multiconductor cable surrounded by a separate solid or braided conductor used to minimize the effects of adjacent electrical circuits.

Shockley diode A four-layer, controlled semiconductor rectifier diode without a base connection, used as a trigger or switching diode. *(See figure on page 136.)*

short See *short circuit.*

short circuit A connection of abnormally low resistance between two points in a circuit. Also called a *short.*

short-circuit admittance parameters See *Y-parameters.*

shorting switch A type of switch arrangement in which one contact is made for a new position before breaking contact with the previous position.

short wave Abbreviated *SW*, radio frequencies in the range from 1.5 to 30 MHz.

shot noise (1) Noise characteristic of all transistors and diodes, generated by the random passage of discrete current carriers across the semiconductor junction. It is proportional to the square root of the applied current. (2) Noise as a result of thermal agitation in vacuum tubes.

shunt (1) A precision resistor placed across the terminals of an ammeter to increase its range. (2) The connecting of a circuit element in parallel with another circuit element or branch.

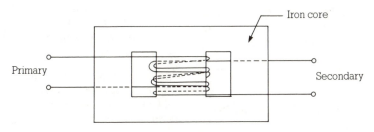

Shell-type transformer

135

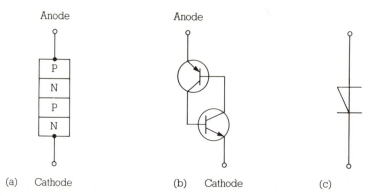

Shockley diode: (a) Internal structure; (b) Equivalent circuit; (c) Symbol

shunt feed The application of a DC voltage to the plate of a vacuum tube, collector of a transistor, or drain of a field effect transistor through a choke coil that is parallel to, and therefore separated from, the signal circuit. Also called *parallel feed*.

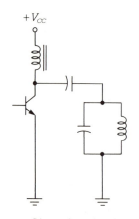

Shunt fed circuit

shunt regulator A device, such as a transistor, placed in parallel with a power source and able to automatically vary its series resistance, which in turn maintains a constant output voltage or current.

shunt-series circuit See *parallel-series circuit*.

shunt-wound generator A DC generator in which the field coils and armature are wound in parallel.

shunt-wound motor A DC motor in which the field coils and armature circuit are connected in parallel.

Si Chemical symbol for *silicon*.

SI Abbreviation for *Système International d'Unités,* the French equivalent of *International System of Units.*

sideband The frequency components of a modulated carrier above and/or below the carrier frequency. The lower sideband is the difference of the carrier and modulating components while the upper sideband is the sum of the carrier and modulating components.

side frequency The frequency of a sideband.

siemen Abbreviated S, the unit of conductance, formerly the mho. Equal to the reciprocal of an impedance or resistance of 1 ohm.

sigma (Σ, or σ) Eighteenth letter of the Greek alphabet, used as the symbol for *conductivity* (σ), *Stefan-Boltzmann constant* (σ), and *summation* (Σ).

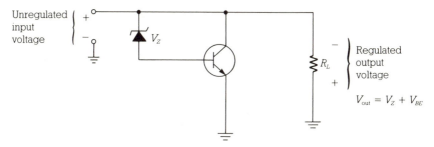

Shunt regulator

$$V_{out} = V_Z + V_{BE}$$

signal   A visible, audible, or other indication that conveys information.

signal generator   A test instrument that can generate an unmodulated or tone-modulated radio or audio frequency sinewave voltage at a known frequency and intensity.

signal strength   The strength of a transmitted radio frequency signal, expressed in microvolts or millivolts per meter of effective receiving antenna height at a particular location.

signal strength meter   A meter, calibrated in decibels or arbitrary S-units, connected to the automatic volume control circuit of a receiver to indicate the strength of a received signal. Also called an *S-meter* or *S-unit meter.*

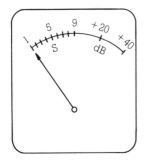

Signal strength meter

signal-to-noise ratio   Abbreviated *S/N ratio,* the ratio, expressed in decibels, of the amplitude of the desired signal at any point to the amplitude of the noise signal at the same point. The peak value is used for pulse noise while the RMS value is used for random noise.

signal tracing   A troubleshooting technique that involves tracing the progress of a signal through each stage of a circuit to locate the faulty stage.

signal voltage   The RMS value of a signal.

silicon   Symbolized by *Si*, a nonmetallic element having semiconductor properties, used in the manufacture of transistors and diodes.

silicon controlled rectifier   Abbreviated *SCR,* a four-layer, three-terminal solid state device that exhibits an open circuit between anode and cathode until a significantly large current flows between the gate and the cathode. It remains as a short circuit between anode and cathode until the anode-to-cathode voltage decreases to zero or goes negative. *(See figure on page 138.)*

137

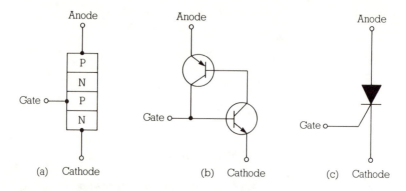

Silicon controlled rectifier: (a) Internal structure; (b) Equivalent circuit; (c) Symbol

silicon controlled switch Abbreviated *SCS*, a four-terminal semiconductor switching device which can be used as a silicon controlled rectifier, complementary silicon controlled rectifier, gate turn-off switch, or transistor.

silicon diode A diode made using silicon as the semiconductor.

silver solder A solder alloy composed of silver, copper, and zinc. It has a melting point lower than silver but higher than lead-tin solder.

SINAD ratio A ratio, expressed in decibels, of the sum of the signal, noise, and distortion to the sum of the noise and distortion at the output of a receiver for a modulated signal input.

sine-cosine generator See *resolver.*

sinewave A fundamental waveform whose amplitude varies as the sine of a linear function of its argument, expressed in either radians, degrees, or time.

single ended Unbalanced, as when one side of a circuit or transmission line is grounded.

single-ended input An amplifier or other circuit in which one side of a two-input connection is grounded.

single sideband Abbreviated *SSB*, pertaining to a modulation system where one of the two sidebands is suppressed or filtered out.

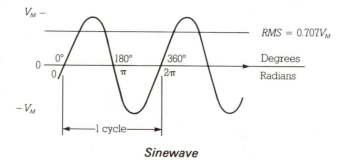

Sinewave

sink Anything that is a power consuming device, such as a load.

skin effect The phenomenon where alternating current concentrates in the surface of a conductor, so that the current distribution through the cross-section of the conductor is not uniform.

sky wave (1) As opposed to a ground wave, a radio wave that travels upward into space and may or may not be returned to earth by reflection from the ionosphere. (2) Ionospheric wave.

slew rate Expressed in volts per microseconds, a measure of an operational amplifier's switching speed, defined as the maximum time rate of change of the output voltage when subjected to a square wave input signal when the closed-loop gain is unity.

slip ring A conducting ring mounted on, but insulated from, a rotating shaft and used in conjunction with a stationary brush to electrically join the fixed and moving parts of a circuit, such as a generator.

slug (1) A copper ring encircling the core of a relay to delay the operation of the relay. (2) In an inductor, a ferromagnetic core whose position

can be varied inside to vary its self-inductance.

small signal A signal level small enough so that a nonlinear device, such as a transistor, can be represented by linear circuit elements.

S-meter See *signal strength meter.*

Smith chart A circular chart on which the real and imaginary parts of any impedance or admittance are plotted on coordinates consisting of two families of orthogonally intersecting circles. The Smith chart allows for the rapid, graphical solution of problems with transmission lines, antennas, and impedance matching. *(See figure on page 140.)*

SNR (or S/N ratio) Abbreviation for *signal-to-noise ratio.*

socket Designated by the letter *J* on schematic diagrams. See *jack.*

solar cell A photovoltaic silicon cell that converts light directly into electrical energy.

solder (1) An alloy that can be melted at low temperatures, used in joining metals which have higher melting points. (2) To join two metals with solder.

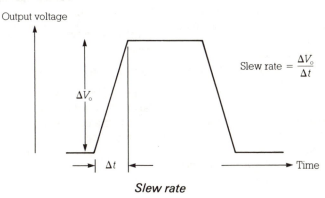

Slew rate

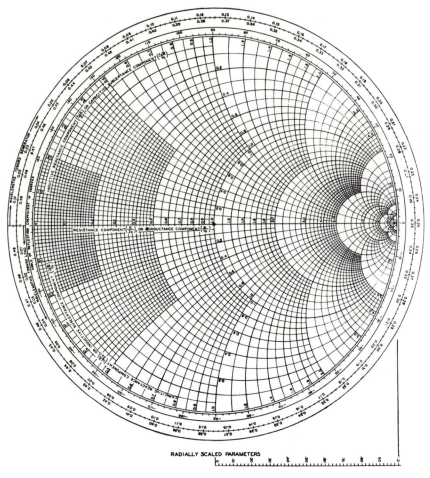

RADIALLY SCALED PARAMETERS

Smith chart

solenoid Symbolized by the letter *K* on schematic diagrams, a coil that surrounds a movable iron core. When the coil is energized by a current, the core is pulled to a different position with respect to the coil. To produce rotary movement, the core is stationary while the coil is movable.

solid-state Pertaining to a circuit, device, or system that depends on a combination of electrical, magnetic, or optical phenomena within a solid that is usually a crystalline semiconductor material.

sonic frequency See *audio frequency.*

sound See *audio frequency.*

source (1) Abbreviated *S*, the terminal of a field effect transistor, comparable to the emitter of a bipolar transistor. (2) A circuit or device which supplies voltage or current to a load.

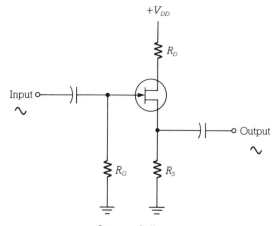

Source follower

source follower amplifier A common-drain FET amplifier having a maximum voltage gain of unity while the output signal is in phase with the input. See *common-drain amplifier.*

source impedance The impedance presented by a source to a load.

spaghetti Insulated tubing used over bare wires.

SPDT Abbreviation for single pole, double throw. A part of a switch having one wiper with the ability to switch an electrical current to one of two paths.

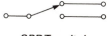

SPDT switch

speaker Abbreviated *SPKR*, an electroacoustic device that converts audio-frequency electric power into acoustic power, radiating it into the air in the form of vibrations, or sound-waves. Also called *loudspeaker.*

specific resistance See *resistivity.*

spectrum All the frequencies used for a particular purpose.

spectrum analyzer A test instrument using a cathode-ray tube that measures and displays the intensities of the frequency components of a complex waveform throughout the frequency range of the waveform.

speech amplifier An audio-frequency amplifier specifically designed to amplify speech frequencies, which are generally in the 300 Hz–3 kHz range.

speech clipper A clipper circuit that limits the peaks of speech-frequency signals for increasing the average percent modulation of a transmitted signal.

speech compressor A device which eliminates certain speech-frequency signals to reduce its bandwidth without affecting intelligibility.

spike A transient of short duration whose intensity well exceeds the signal's average intensity.

SPKR (or spkr) Abbreviation for *speaker.*

141

split-phase motor A single phase induction motor that has an auxiliary winding connected in parallel with, but physically displaced from, the main winding to produce the required rotating magnetic field for starting. The auxiliary circuit is normally opened when the motor reaches a predetermined speed.

split-series motor A DC, series connected motor that has one field winding for each direction of rotation.

spot size The cross-section of an electron beam at the screen of a cathode-ray tube.

SPST Abbreviation for single pole, single throw. A part of a switch having one wiper with the ability to switch an electrical current on and off.

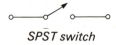

SPST switch

square wave A fundamental waveform that alternately equals two fixed values for equal lengths of time.

squeal In a radio receiver or audio amplifier, a high-pitched tone heard together with the desired signal.

squegg To oscillate and cut off for alternate periods of time.

squegging oscillator See *blocking oscillator.*

squelch To automatically quiet a receiver by reducing its gain in response to the absence of an input signal.

squirrel-cage winding A permanently short-circuited winding, usually insulated, that has its conductors uniformly distributed around the edge of the rotor and which is joined by continuous end rings.

SSB Abbreviation for *single sideband.*

stability Freedom from undesired variations or oscillations.

stabilized feedback See *negative feedback.*

stage One of the functional parts into which a circuit can be conveniently divided, containing one or more active devices.

stagger tuned Adjustment of successively tuned amplifier or filter stages to slightly different frequencies in order to widen the overall frequency response of the circuit.

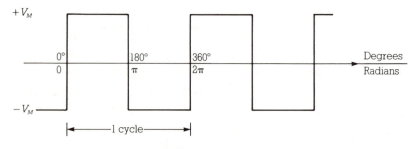

Square wave

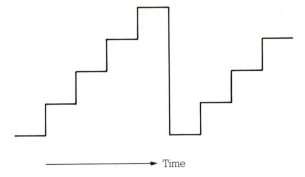

Staircase waveform

staircase A waveform whose intensity increases in discrete, uniform steps thus giving the appearance of a staircase.

standing wave A wave in which the ratio of an instantaneous value at one point to that at another point does not vary with time, which is produced by two waves of the same frequency traveling in opposite directions.

standing wave ratio Abbreviated *SWR*. See *voltage standing wave ratio.*

standoff An insulator that supports a conductor at a distance from the surface on which the insulator is mounted.

star connection See *wye connection.*

state-variable filter An active filter using operational amplifiers as summing amplifiers and integrators to produce simultaneous low-pass, high-pass, bandpass, and notch filter responses. Also called a *universal filter.*

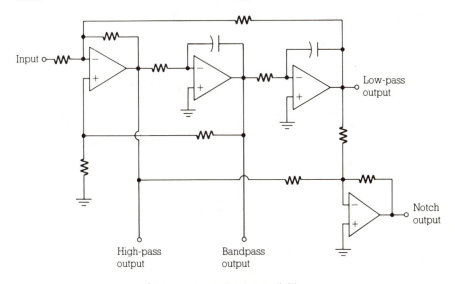

State-variable (universal) filter

static (1) A hissing, crackling, or other sharp sound caused by electric storms or improperly operating electrical devices. The sound is usually heard in radio receivers or appears as white snow on a television picture. (2) Without motion or change.

static electricity The transfer of static charge from one object to another either by direct contact or by a spark that bridges an air gap between the objects.

stator (1) The portion of a rotating machine that contains the stationary parts of the magnetic circuits and their associated windings. (2) The stationary set of parallel plates in a variable capacitor.

steady-state The condition in which circuit values remain essentially constant after initial transients have disappeared.

Stefan-Boltzmann constant Symbolized by the Greek letter σ, equal to 5.672×10^{-8} joule per second-meter²-kelvin⁴.

step-down Pertaining to a transformer whose secondary voltage is less than its primary voltage.

step function A mathematical function whose value is zero before a certain instant of time, and equal to a constant, nonzero value after that instant.

stepper motor A motor that rotates in short, uniform, angular movements rather than continuously.

stepping relay A relay whose contact arm may rotate through 360 degrees, but not in one operation.

step-up Pertaining to a transformer whose secondary voltage is greater than its primary voltage.

stop band The frequency range in which a filter has high attenuation. Also called *rejection band.*

stray capacitance Undesirable capacitance between circuit wires, wires and the chassis, or components and the chassis of electronic equipment.

stray field Leakage magnetic flux that spreads outward from a coil and does no useful work.

stub A section of transmission line, the far end either open or shorted, connected in parallel with a transmission line at a particular point to match the impedance of the line to that of a load.

subcarrier A carrier signal that is applied as a modulating signal to modulate another carrier.

subharmonic A sinewave whose frequency is an integral submultiple of a fundamental frequency. A fifth subharmonic is one-fifth the fundamental frequency.

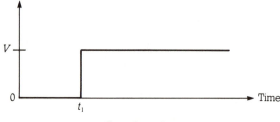

Step function

Subsidiary Communications Authorization Abbreviated *SCA*, an FCC regulation that authorizes the multiplexing of FM broadcasts to provide background music services.

sum frequency In a frequency converter or mixer, the output frequency equal to the sum of its two input frequencies.

summation Symbolized by the Greek letter Σ, the mathematical operation where all terms of a particular group or series are added together.

summing amplifier Also called a *summer,* an operational amplifier circuit which takes the instantaneous algebraic sum of two or more input signals.

S-unit An arbitrary unit of signal strength, sometimes used along with decibels on the scale of a signal strength meter. Frequently, one S-unit equals 6 dB.

S-unit meter See *signal strength meter.*

superhet Abbreviation or contraction for *superheterodyne.*

superheterodyne A type of radio receiver in which modulated input carrier frequencies are converted or mixed to a fixed, or intermediate, fre-

quency by a local oscillator, after which amplification takes place. Receiver tuning is done by varying the frequency of the local oscillator. Also called *superhet.*

super high frequency Abbreviated *SHF,* any frequency in the region from 3 to 30 GHz.

superposition theorem A theorem stating that the response of a linear system to multiple inputs acting simultaneously is equal to the sum of the responses of the inputs taken individually.

supply voltage The voltage obtained from a power supply for operation of a circuit or device. Called *rail voltage* in Britain.

suppressor grid A vacuum tube element positioned between the screen grid and the plate, which is kept at a fixed potential less than either the screen grid or plate to prevent the transfer of secondary electrons from the screen grid to the plate. Also called the *third grid* or G_3.

surge impedance See *characteristic impedance.*

susceptance (1) Symbolized by the letter B, the imaginary component of admittance. (2) The reciprocal of reactance.

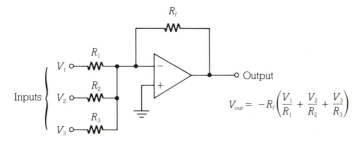

Summing amplifier

susceptibility Symbolized by the Greek letter χ, the ratio of the magnetization of a magnetic material to the inducing magnetic field strength.

SW (1) Abbreviation for *short wave.* (2) Abbreviation for *switch.*

swamping resistor An unbypassed resistor placed in the emitter lead of a bipolar transistor circuit to minimize the effect of temperature on the AC base-emitter junction resistance.

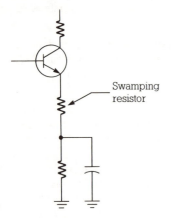

Swamping resistor

Swamping resistor

sweep (1) In a cathode-ray tube, the horizontal movement of the electron beam. (2) Anything that changes steadily with time.

sweep generator A test instrument that generates either an audio-fre-

quency or radio-frequency signal that varies back and forth through a given frequency range at a constant rate.

SWG Abbreviation for Standard Wire Gage.

swing The total variation in the frequency or amplitude of a quantity, such as the voltage swing.

swinging choke An iron-core inductor designed so that its inductance decreases as the current through it increases. Consequently, the inductance swings between its maximum and minimum values as the current changes. It is used as part of a choke input filter in power supplies to improve voltage regulation.

switch (1) Abbreviated *S* or *SW,* a manual or mechanically actuated device for making or breaking electrical paths. (2) To control by means of a switch. (3) To change from one state to another.

SWR Abbreviation for *standing wave ratio.*

sync Synchronize, or synchronization.

synchronized sweep A sweep voltage controlled by an AC voltage so that the forward and return traces displayed on a cathode-ray tube are exactly superimposed and appear as a single trace.

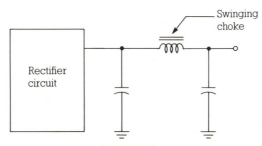

Swinging choke

synchronous   In step or in phase, as applied to two or more circuits, devices, or machines.

synchronous converter   Normally used to change AC power to DC power, a converter in which the motor and generator windings are combined on one armature and excited by a single magnetic field.

synchronous generator   See *alternator.*

synchronous inverter   A dynamotor.

synchronous machine   An AC machine whose average speed is proportional to the frequency of the applied or generated voltage.

synchronous motor   A synchronous machine that transforms AC power into mechanical power by using DC-excited field magnets.

synchronous speed   A speed related to the frequency of the AC power line and the number of poles in a synchronous machine. For N poles and a given line frequency f, the number of revolutions per minute is equal to rpm $= 120 f/N$.

synchro resolver   See *resolver.*

system   A group of components, or several pieces of equipment, interconnected or integrated together to perform a specific function.

system noise   The output from a system with no input signal.

147

t Symbol for time.

T (1) Unit symbol for *tesla*. (2) Prefix symbol for *tera* (10^{12}). (2) Symbol for *period* of oscillation, *time constant,* or temperature.

tail The trailing edge of a pulse.

tank circuit A resonant LC circuit.

tap An electrical connection made to a resistor or inductor coil at a given point along its length.

taper Relating to the angular position of a shaft of a potentiometer or rheostat whose variation in resistance may be either a linear or a logarithmic function of the shaft rotation angle.

tau (τ) The nineteenth letter of the Greek alphabet, used as the symbol for *time constant,* or *transmittance.*

TC Abbreviation for *temperature compensated,* or *temperature coefficient.*

Tchebycheff French spelling for *Chebyshev.*

Tchebycheff filter See *Chebyshev filter.*

TCXO Abbreviation for *temperature compensated crystal oscillator.*

TDR Abbreviation for *time domain reflectometry.*

TE Abbreviation for *transverse electric.* See *transverse electric wave.*

tee-network See *T-network.*

telemetering (1) The transmitting of the value of a measured quantity to a remote point by radio or over wires. (2) Also called telemetry, the complete measuring, transmitting, and receiving apparatus for indicating and recording the value of a quantity at a distance.

telephony The radio transmission of speech and sounds to a distant point for communication purposes.

television Abbreviated *TV,* a system that converts a series of visual images into electrical signals to be transmitted by radio waves to a distant point.

television broadcast band The radio frequency spectrum in which 6-MHz wide channels are used for broadcasting TV signals in the U.S. The VHF channel frequencies range from 54 to 74 MHz (TV channels 2 through 4), 76 to 88 MHz (TV channels 5 and 6), 174 to 216 MHz (TV channels 7 through 13). The UHF channel frequencies (TV channels 14 through 83) range from 470 to 890 MHz.

TEM Abbreviation for *transverse electromagnetic.* See *transverse electromagnetic wave.*

tempco Acronym for *temperature coefficient.*

temperature coefficient Abbreviated *TC,* or *tempco,* the amount of change in a value of a parameter of a device or component per degree change in temperature.

temperature compensated Abbreviated *TC,* when some circuit characteristic is made independent of temperature changes.

temperature compensated crystal oscillator Abbreviated *TCXO,* a crystal-controlled oscillator in which temperature-compensation networks have been adjusted to match the crystal characteristics so that frequency stability is maintained over a wide temperature range.

tera Abbreviated *T,* prefix for 10^{12}.

terminal A point at which a component or network can be connected to another component or network.

termination The load that is connected to the output of a device or transmission line.

tesla Abbreviated *T,* a unit of magnetic induction equal to 1 weber per square meter. Named after Nikola Tesla, an American inventor.

Tesla, Nikola An American inventor (1856–1943), born in Croatia (now Yugoslavia), who pioneered the development of radio and high-tension electricity. His specific developments included the Tesla induction motor, Tesla coil, and the AC generator.

Tesla coil An induction coil having a high-frequency spark gap that induces a secondary voltage of such intensity to cause an intense, high-frequency discharge.

test lead A flexible, insulated lead, usually with a probe at one end, used for connecting instruments to circuits temporarily for test measurements.

tetrode A vacuum tube that has four elements: cathode, control grid, screen grid, and plate.

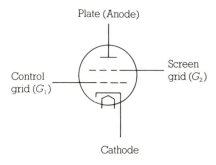

Tetrode

THD Abbreviation for *total harmonic distortion.*

thermal cutout A heat-sensitive switch that automatically opens the circuit of an electric motor or other device when the operating temperature exceeds a safe value.

thermal drift   A departure of a value, such as the frequency of an oscillator, caused by internal heating of equipment during normal operation or by changes in environmental temperature.

thermal noise   Random noise as a result of the thermal agitation of atoms in carbon resistors and semiconductor materials. Also called Johnson noise.

thermal resistor   See *thermistor.*

thermal runaway   A condition in transistors in which heating of the collector-emitter junction causes a rise in collector current, which in turn causes more heating. At its limit, the junction temperature rapidly approaches its maximum rating, beyond which the transistor will be destroyed. Also called *runaway.*

thermionic   Pertaining to the emission of electrons as a result of heat.

thermistor   Contraction for *thermal resistor,* a device that makes use of the change in the resistivity of a semiconductor with a change in temperature. A thermistor has a high negative temperature coefficient of resistance so that its resistance decreases as the temperature increases. It is used chiefly in critical circuits to compen-

sate for temperature variations in other components.

thermocouple   A two-terminal device, based on the Seebeck effect, which is composed of two dissimilar metals that produce a voltage across this junction that is linearly proportional to the temperature of the junction.

thermoelectric effect   See *Seebeck effect.*

thermorelay   See *thermostat.*

thermostat   A two-terminal switch that opens or closes a circuit when the temperature changes from a preset value or range of values. Also called a *thermorelay.*

theta (θ)   Eighth letter of the Greek alphabet, used as a symbol for temperature or angle.

Thevenin's theorem   In a linear network, the current through an impedance connected across a pair of terminals A and B is the same as if the impedance were connected to an ideal voltage source, equal to the open circuit voltage between terminals A and B, and in series with an impedance equal to the impedance between terminals A and B when all independent voltage sources are short-circuited.

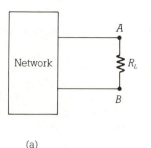

(a)

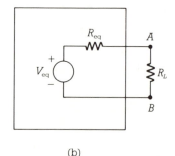

(b)

Thevenin's theorem: (a) General circuit; (b) Thevenin equivalent circuit

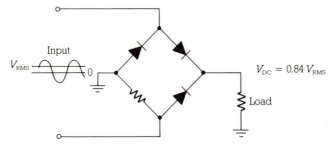

Three-quarter bridge

third grid   See *suppressor grid.*

Thompson bridge   See *Kelvin bridge.*

3-dB frequency   See *cutoff frequency.*

three-phase circuit   A circuit powered by three AC voltages whose phases differ by 120 degrees.

three-quarter bridge   A bridge circuit in which one of the diode rectifiers is replaced by a resistor. The average output voltage is 0.84 times the input RMS voltage.

tickler   A small coil connected in series with either the plate circuit of a vacuum tube or the collector circuit of a transistor, which inductively couples the grid- or base-circuit coil to provide positive feedback in regenerative detector and oscillator circuits.

tie point   An insulated terminal to which two or more wires are connected.

time base   (1) A voltage generated by the sweep circuit of a cathode-ray tube so that its trace is linear with respect to time. (2) A stable, crystal-controlled oscillator in digital logic circuits.

time constant   Symbolized by T or the Greek letter τ, the time required for a circuit parameter, such as voltage or current, to either rise to $1 - 1/e$ (approximately 63 percent) or decay to $1/e$ (approximately 37 percent) of its steady-state value after a step-function change in excitation. In a resistor-capacitor network, equal to the product of resistance and capacitance. In a resistor-inductor network, equal to the inductance divided by the resistance. In an inductor-capacitor network, equal to the product of inductance and capacitance.

time delay   (1) The time required for a signal to travel between two points. (2) In a monostable multivibrator, the time that the output is in its unstable state after triggering. Also called *output pulse width.*

time domain reflectometry
Abbreviated *TDR*, an instrument that measures the electrical characteristics of wideband transmission systems, such as transmission lines, by feeding in a step voltage and displaying the superimposed reflected signals on a cathode-ray tube. The display then gives the nature and location of each pulse reflecting a discontinuity.

timer (1) A special clock mechanism or motor-operated device used to perform switching operations at predetermined time intervals. (2) A class of integrated circuit devices which can be wired to function as a monostable multivibrator.

timer motor A synchronous clock motor with a specific output speed determined by the number of poles in the stator as well as the associated gear train.

tip jack A small single-hole jack for a single-pin contact plug. Also called a *pup jack*.

TM Abbreviation for *transverse magnetic*. See *transverse magnetic wave*.

T-network A three-branch network resembling the letter *T*. Its electrical characteristics are identical to a star, or Wye-network. Also spelled *tee-network*.

toggle switch A two-position, snap-action switch operated by a projecting lever to open or close circuit paths.

TO package A packaging style used for some transistors, integrated circuits, and other semiconductor devices in which the leads are generally arranged in a circular pattern and project from the base parallel to the axis of the device.

toroid (1) A surface having the shape of a doughnut with a center hole, which is generated by revolving a plane, closed curve about a line that lies in its plane but has no point in common with it. (2) A coil wound upon a toroid core, which results in a highly concentrated magnetic field within itself while having a minimum amount of flux leakage (external field).

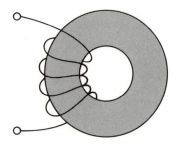

Toroid coil

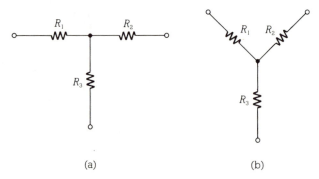

(a)

(b)

T-network: (a) Standard arrangement; (b) Equivalent Y-network

total harmonic distortion Abbreviated *THD*, either expressed as a percentage or in decibels, a measure of nonlinearity which is the ratio of the power of the components of the higher-order harmonics of the output signal of a system, such as an amplifier, to the fundamental component of the output signal of a system, such as an amplifier.

Touch-Tone® A registered trademark of AT&T, which is a telephone pushbutton dialing system using two of seven possible non-harmonically related, audio tone frequencies for the numbers 0 through 9 and the symbols # (pound) and * (star). The audio tones are divided into two frequency range groups so that any number or symbol uses only one frequency from each group in the matrix arrangement at the bottom of the page.

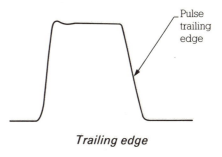

Touch-Tone® telephone dialing pad

T-pad A pad composed of resistance elements arranged in a T-network.

TR (or T/R) Abbreviation for transmit/receive.

trace The pattern, usually a single line, on the screen of a cathode-ray tube.

tracking (1) The maintaining of proper frequency relationships between tuned circuits in a receiver so that they accurately follow the tuned frequency over its entire range. (2) In power supplies, the interconnection such that one power supply serves to control all units in series operation so that the output voltage of the slave follows the output voltage of the master.

trailing edge The major portion of the decay of a pulse waveform.

Trailing edge

train A sequence of pulses or waves.

trans Abbreviation for *transmitter.*

Low-Tone Group	High-Tone Group		
	1 209 Hz	**1 336 Hz**	**1 477 Hz**
697 Hz	1	2	3
770 Hz	4	5	6
852 Hz	7	8	9
941 Hz	*	0	#

153

transadmittance A conjunction of *transfer admittance.*

transconductance Symbolized by g_m, a parameter of either a vacuum tube equal to the change in plate current to the corresponding change in grid voltage, or of a field effect transistor equal to a change in drain current to the corresponding change in gate voltage. Also called *mutual conductance.*

transducer Any device by which energy is transferred from one system to another, particularly when the input and output energy are of different forms.

transfer admittance The reciprocal of transfer impedance. Also called *transadmittance.*

transfer function In an electrical or mechanical network, a mathematical function that relates the response of one pair of terminals to a source applied at another pair of terminals. In most cases, a transfer function relates the output response of a network to the input as a function of time or frequency.

transfer impedance An impedance equal to the ratio of the voltage applied between two terminals of a circuit to the resultant current at another pair of terminals in the circuit.

transfer resistor See *transistor.*

transformer A device for transferring AC current from one circuit to another by magnetic induction. Each circuit is connected to a coil wound around a magnetic core so that the power in both windings is the same. However, the output voltage depends upon the turns ratio of the transformer.

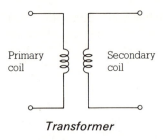

Transformer

transformer oil A high-quality insulating oil in which windings of transformers are immersed to provide high dielectric strength, high insulation resistance, high flash point, and freedom from moisture and oxidation.

transient A pulse, damped oscillation, or other temporary behavior occurring in a system prior to reaching its steady-state condition.

transient oscillation A momentary oscillation that occurs in a circuit during switching.

transient overshoot See *overshoot.*

transient response The behavior of a circuit, prior to reaching a steady-state condition, in response to a sudden change in an input quantity, such as a step voltage.

transient suppressor A device, such as a Zener diode or varistor, used to protect a circuit from destructive voltage surges.

transistor Contraction of *transfer resistor* and designated by the letter *Q* on schematic diagrams, a solid-state, current-gain device having a collector, base and emitter terminals. Made using both p- and n-type materials, transistors are available in either PNP or NPN configurations. Also called a *bipolar transistor.* See also *NPN transistor* and *PNP transistor.*

transistorized DC motor (1) A conventional AC motor driven by a transistorized DC/AC converter. (2) A DC motor in which transistors replace the commutator for commutating the current.

transition factor See *reflection coefficient.*

transition frequency See *cross-over frequency.*

transmission line A waveguide, coaxial cable, or other system of conductors used to efficiently transfer signal energy from one location to another.

transmission loss (1) The ratio, usually expressed in decibels, of the power at one point in a transmission system to the power at some point farther along the line. (2) The actual power that is lost in transmitting a signal from one point to another through a medium or along a line. Also called *line loss.*

transmittance Symbolized by the Greek letter τ, the ratio of the radiant power transmitted through a material to the incident radiant power.

transmitter Abbreviated *trans,* or *xmtr,* a device used for generating and amplifying an RF carrier signal, modulating the carrier, and feeding the modulated signal to an antenna for radiation into space as electromagnetic waves. Also called a *rig.*

transponder A radio device that receives an incoming signal and automatically retransmits it on the same or on a different frequency.

transrectification Rectification that occurs in one circuit when an AC voltage is applied to another circuit.

transresistance The reciprocal of transconductance.

transverse electric wave Abbreviated *TE wave,* an electromagnetic wave in which the electric field vector is everywhere perpendicular to the direction of propagation. Called *H-wave* in Britain.

transverse electromagnetic wave Abbreviated *TEM wave,* an electromagnetic wave in which both the electric and magnetic field vectors are everywhere perpendicular to the direction of propagation.

transverse magnetic wave Abbreviated *TM wave,* an electromagnetic wave in which the magnetic field vector is everywhere perpendicular to the direction of propagation. Called *E-wave* in Britain.

trap A filter or tuned resonant circuit used to suppress an unwanted signal. Also called a *rejector.*

trapezoidal wave An oscilloscope display in which an amplitude-modulated carrier is connected to the horizontal amplifier. By analyzing the resulting trapezoid-shaped waveform, the percentage of modulation can be determined. *(See figure on page 156.)*

traveling wave A wave, propagating through a medium, whose amplitude is a function of time and position.

treble (1) High audio frequencies such as those handled by a tweeter in a sound system. (2) Audio frequencies above middle C (256 Hz).

tree A set of connected branches without meshes.

tremolo The amplitude modulation of an audio tone resulting in a warbling or fluctuating tone.

triac

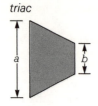

$$\% \text{ Modulation} = \frac{a - b}{a + b} \times 100$$

Less than
100% modulation

100%
Modulation

Greater than
100% modulation

Trapezoidal wave pattern

triac   Acronym for triode AC semiconductor switch, a five-layer semiconductor device that is equivalent to two silicon controlled rectifiers in antiparallel, having a common gate and two anode terminals. It provides switching action for either polarity of applied voltage and can be con-trolled in either polarity by the gate terminal.

triangle wave   A waveform that is a repeating ramp function with equal positive and negative slopes.

trickle charge   The continuous charging of a storage battery at a low rate to keep it fully charged.

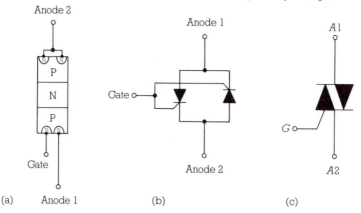

Triac: (a) Internal structure; (b) Equivalent circuit; (c) Symbol

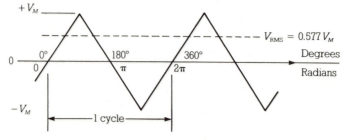

Triangle waveform

trigger (1) To cause, by means of one circuit, action to start in another circuit that functions for a given length of time under its own control. (2) A pulse that starts an action in another circuit.

trigger diode See *diac.*

trigger pulse A pulse used for triggering. Also called a *tripping pulse.*

trim To make a fine adjustment of a circuit or circuit element. Also called *tweak.*

trimmer A small adjustable-value circuit element, generally a capacitor, connected in conjunction with a circuit element of the same kind in such a way that its adjustment changes the combination of the two elements to a given value.

triode A vacuum tube that has three elements: cathode, control grid, and plate.

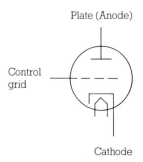

Plate (Anode)

Control grid

Cathode

Triode

trip coil A coil that opens a circuit breaker or other protective device when the current through the coil exceeds a predetermined value.

tripping pulse See *trigger pulse.*

trip value The value of current, power, or voltage at which a polarized relay will transfer from one contact to another.

Tschebycheff German spelling for *Chebyshev.*

T²TL Abbreviation for transistor-transistor logic.

TTL Abbreviation for transistor-transistor logic.

TTY Abbreviation for teletypewriter.

tube See *electron tube.*

tube shield A Faraday shield placed around a vacuum tube.

tubular capacitor A capacitor that is in the form of a cylinder with axial leads.

tune To adjust for resonance at a desired frequency.

tuned amplifier An amplifier in which the load is a tuned circuit. Consequently, the load impedance and amplifier gain vary with frequency.

tuned circuit An LC circuit which can be adjusted for resonance at a desired frequency.

tuned radio-frequency receiver A radio receiver whose amplifier stages are tuned to resonate at the carrier frequency of the received signal by a ganged variable capacitor. The amplified signal at the carrier frequency is demodulated and is further amplified and fed to a speaker or headphone.

tuned-reed frequency meter See *vibrating-reed frequency meter.*

tuning coil A variable-inductance coil used for adjusting the frequency of an oscillator or resonant circuit.

tuning core A ferrite core that is moved in and out of a coil or transformer to vary the inductance.

tuning range The frequency range over which a receiver or other piece of equipment can be adjusted.

tuning stub A short length of transmission line, usually shorted at the free end, that is connected to a transmission line for impedance matching purposes.

tunnel diode A two-terminal semiconductor device having in the region of low forward bias a negative resistance characteristic for use as an oscillator or amplifier. Also called an *Esaki diode.*

turns ratio Symbolized by *N.* In a transformer, the ratio of the number of turns of the primary winding to the number of turns of the secondary winding.

TV Abbreviation for *television.*

TVI Abbreviation for television interference.

tweak To make a small refining adjustment on a circuit after it has been installed. Also called *trim.*

tweeter A loudspeaker that handles only the higher audio frequencies, generally above 3 kHz.

twin lead A flat transmission line that has two parallel conductors separated by insulated material, and whose impedance is 300 ohms. Also called a *balanced transmission line.*

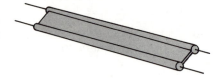

Twin-lead

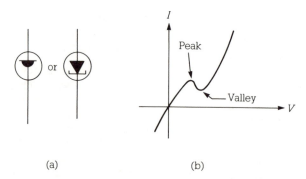

(a)   (b)

Tunnel (Esaki) diode: (a) Symbol; (b) V-I *characteristic*

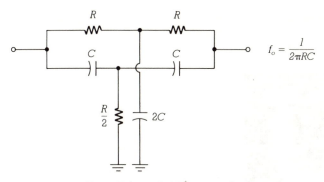

$$f_o = \frac{1}{2\pi RC}$$

Twin-T (parallel-T) network

twin-T network   A network that consists of two T-networks in parallel, with the capacitor and resistor portions interchanged in one of them. By itself, it has the response of a notch filter, but when placed in the feedback path of an operational amplifier, it has the response of a bandpass filter. Also called a *parallel-T network*.

twisted pair   A cable composed of two small, insulated conductors twisted together without a common covering.

two-phase   (1) Having a phase difference of one-quarter cycle, or 90 degrees. Also called *quarter-phase*. (2) Quadrature.

U Letter designation for an *integrated circuit* on a schematic diagram.

UHF (or uhf) Abbreviation for *ultrahigh frequency.*

UJT Abbreviation for *unijunction transistor.*

UL Abbreviation for *Underwriters' Laboratory.*

ultrahigh frequency Abbreviated *UHF*, any frequency in the region from 300 MHz to 3 GHz.

ultrasonic Having a frequency above the audio-frequency range.

ultrasonic frequency Any frequency in the region above the audible range, roughly above 20 kHz.

ultraviolet Abbreviated *UV*, pertaining to electromagnetic radiation at those wavelengths from 10 to 380 nanometers, which are beyond the visible (violet) end of the light spectrum.

unbalanced (1) A circuit which generally has one side grounded. (2) Differential mutual impedance or mutual admittance between two circuits which ideally would have no coupling.

unbalanced circuit A circuit which is not symmetrical with respect to a common reference point, usually ground.

unbalanced line A transmission line in which the voltages with respect to ground on the two conductors are not equal.

unbalanced output An output in which one of the two input terminals is essentially at ground.

undamped natural frequency The frequency at which a system will oscillate without damping after the system has been momentarily disturbed.

undamped oscillations Oscillations that have a constant amplitude indefinitely.

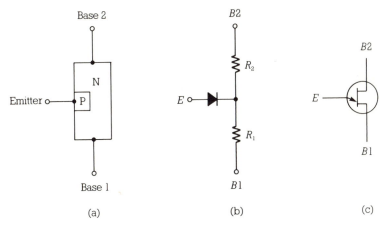

Unijunction transistor: (a) Internal structure; (b) Equivalent circuit; (c) Symbol

underdamped A low level of damping that is not able to prevent a circuit from oscillating after the application of a sharply rising input signal.

Underwriters' Laboratory Abbreviated *UL*, a corporation that establishes safety standards on electrical appliances, components, and equipment.

ungrounded Not grounded; i.e., not connected directly to ground potential except through high impedance devices.

unijunction transistor Abbreviated *UJT*, a three-terminal semiconductor device with a single emitter lead and two base leads, used primarily as a switching device.

unipolar transistor See *junction field effect transistor.*

unity coupling Ideal magnetic coupling between two inductors so that all the magnetic flux produced by the primary winding passes through the entire secondary winding.

unity-gain bandwidth (1) The frequency at which the frequency response of a system decreases to unity voltage gain, or 0 dB. (2) A measure of the gain-frequency response of an amplifier.

unity-gain crossover frequency The frequency at which the open-loop voltage gain of an amplifier is unity (0 dB).

unity power factor A power factor of 1.0 only when the AC voltage and current are in phase. Consequently, the circuit cannot have reactive components.

universal filter See *state-variable filter.*

universal motor An electric motor capable of operating by either a single phase AC or DC current at essentially the same speed and torque. Also called a *series motor.*

untuned Not resonant at any of the frequencies of interest.

upconverter A circuit that changes an incoming carrier frequency to a higher frequency.

upper sideband Abbreviated *USB*, the frequency or group of frequencies generated above the carrier fre-quency in an amplitude modulated system.

USB Abbreviation for *upper sideband*.

UV Abbreviation for *ultraviolet*.

v Mathematical symbol for *AC voltage.*

V (1) Unit symbol for *volt,* or *potential difference.* (2) Mathematical symbol for *DC voltage.* (3) Letter designation for a *vacuum tube* on a schematic diagram.

VA Abbreviation and unit symbol for *volt-ampere.*

vacuum capacitor A capacitor whose plates are concentric cylinders enclosed in a vacuum so that its breakdown voltage is increased.

vacuum switch A switch whose contacts are enclosed in a vacuum, usually to minimize sparking between contacts.

vacuum tube See *electron tube.*

vacuum-tube voltmeter Abbreviated *VTVM,* a high impedance electronic voltmeter in which the measured voltage is amplified using a vacuum tube prior to being read by a meter.

valley current The minimum current in a negative resistance device, such as a unijunction transistor or tunnel diode.

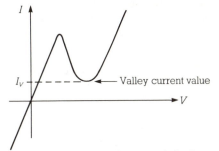

Valley current of a tunnel diode

valve A British term for a *vacuum tube* or *electron tube.*

VAR (or var) (1) Abbreviation and unit symbol for *volt-amperes reactive.* (2) The unit of reactive power.

varactor A two-terminal solid-state device whose capacitance can be varied by changing the voltage across it. Also called a *variable capacitance diode, varicap,* or *voltage controlled capacitor.*

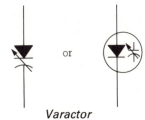

Varactor

variable capacitance diode See *varactor.*

variable capacitor A capacitor whose capacitance can be easily changed over a given range either by varying the effective area of its plates or by changing the distance between them.

variable frequency oscillator Abbreviated *VFO,* an oscillator whose output frequency can be varied over a given range.

variable-mu tube See *remote cutoff tube.*

variable resistor A resistor whose resistance can be easily changed over a given range. See also *potentiometer* and *rheostat.*

Variac A trade name of the General Electric Company, applied to an autotransformer consisting of a fixed toroidal winding and a rotating carbon wiper whose output voltage is adjustable from 0 to 117 percent of the power line voltage.

varicap See *varactor.*

varistor A two-terminal, solid-state device whose nonlinear resistance decreases as the applied voltage increases.

varmeter A meter that is used for measuring reactive power, usually in units of vars or kilovars. Also called a *reactive volt-ampere meter.*

V-band The radio-frequency band from 46 to 56 GHz.

VCO (or vco) Abbreviation for *voltage-controlled oscillator.*

VCXO Abbreviation for *voltage-controlled crystal oscillator.*

vector Any quantity that has both magnitude and direction, usually represented by a boldfaced letter. Declining in use but still current, a given vector A is also represented by the symbol $\overrightarrow{A}$.

velocity factor The ratio of the velocity of wave propagation in a coaxial cable to the velocity wave propagation in free space.

vernier A device, control, or scale markings used to obtain a fine adjustment to increase the precision of a measurement.

vernier dial A tuning dial mechanism where each complete rotation of the control knob causes a fraction of a revolution of the main control shaft to permit fine, accurate, and repeatable adjustments.

vertical amplifier An amplifier for signals intended to produce a vertical deflection on the screen of an oscilloscope. Also called a *Y-axis amplifier.*

vertical metal oxide semiconductor Abbreviated *VMOS,* a metal oxide semiconductor device in which the gate lead is constructed in the form of a V-shaped channel between the source and drain regions of the semiconductor material. This construction allows for higher current flow between source and drain as well as for fast pulse response.

vertical polarization Pertains to radio waves so that the electric (E) field vector is vertical while the magnetic (H) field vector is horizontal.

very high frequency Abbreviated *VHF,* any frequency in the region from 30 to 300 MHz.

very low frequency Abbreviated *VLF,* any frequency in the region from 10 to 30 kHz.

vestigial sideband An amplitude-modulated transmission whereby a portion of one of the two sidebands has been largely suppressed by a filter that has a gradual rolloff in the vicinity of the carrier frequency. The remaining sideband is transmitted without suppression. Also called an *asymmetrical sideband.*

V/F Abbreviation for voltage-to-frequency.

V/F converter See *voltage-controlled oscillator.*

VFO (or vfo) Abbreviation for *variable frequency oscillator.*

VHF (or vhf) Abbreviation for *very high frequency.*

vibrating-reed frequency meter A mechanical frequency meter containing a row of tempered steel rods, or reeds, each of which has a different natural frequency, and which are electromagnetically vibrated in synchronization with the frequency of the supplied AC voltage. The reed that vibrates the most is the nearest to the AC frequency being measured. Also called a *Frahm frequency meter,* or *tuned-reed frequency meter.*

vibrating-reed relay A relay in which the application of an alternating or self-interrupted voltage to its coil produces an alternating or pulsating magnetic field that causes an internal reed to vibrate.

video Pertaining to the name of circuits or parts which carry picture signals.

video amplifier A wideband amplifier capable of amplifying video signals in television receivers, radar, or computer displays.

video frequency Picture-carrying signals in the 15 Hz–5 MHz range.

VLF (or vlf) Abbreviation for *very low frequency.*

VMOS Acronym for *vertical metal oxide semiconductor.*

vogad Acronym for *voice-operated gain-adjusting device.*

voice coil An inductive coil attached to the diaphragm of a moving-coil loudspeaker that moves through the air gap between the pole pieces as a result of the interaction of the fixed magnetic field with the varying current flowing through the voice coil.

voice-operated gain-adjusting device (1) Abbreviated *vogad,* a voice-operated circuit used to produce essentially the same output intensity for a wide range of input levels. (2) A compressor or speech processor.

volt (1) Abbreviated V, the unit of electromotive force (EMF) or potential difference. Equal to the electromotive force of 1 joule per coulomb, and to the voltage required to produce a current of 1 ampere flowing through a 1 ohm resistor.

Volta, Alessandro An Italian physicist (1745–1827) who invented the voltaic pile, the first electric cell.

voltage Symbolized by the letters E or V, the electromotive force which causes current to flow through an electrical conductor. Sometimes called *potential.*

voltage-controlled capacitor See *varactor.*

voltage-controlled crystal oscillator Abbreviated *VCXO,* a crystal-controlled oscillator whose output frequency is a function of an applied input voltage.

voltage-controlled oscillator

Abbreviated *VCO*, an oscillator whose output frequency is a function of an applied input voltage. Also called a *voltage-to-frequency (V/F) converter.*

voltage divider

A network consisting of series-connected impedances connected across a voltage, from which one or more voltages can be obtained across any portion of the network.

voltage drop

The difference in voltage between the two terminals of a branch, or nodes of a circuit.

voltage feedback

Amplifier feedback whereby the voltage drop across part of the load impedance is put in series with the input signal voltage.

voltage follower

An operational amplifier circuit with a direct feedback connection that results in unity gain and zero phase shift. It has a very high input impedance and very low output impedance.

voltage gain

(1) Abbreviated A_V, the ratio of the output voltage of a network to the applied input voltage. (2) Expressed in decibels, the voltage gain is

$$A_V = 20 \log \frac{V_{out}}{V_{in}}$$

voltage multiplier

A rectifying circuit that produces a DC voltage whose amplitude is approximately equal to an integer multiple of the peak amplitude of the applied AC voltage.

voltage-range multiplier

See *instrument multiplier.*

voltage ratio

In a transformer, the ratio of the RMS primary to the RMS secondary terminal voltages under specified load conditions.

voltage reference

In a power supply, a highly regulated DC voltage used as a standard to which the power supply's output voltage is continuously compared to maintain voltage regulation.

voltage regulator

A discrete or monolithic integrated circuit network that maintains its output voltage at a predetermined value independent of changes in the load current or input voltage, as long as the input is greater than the predetermined output voltage.

voltage regulator tube

Abbreviated *VR* tube, a cold-cathode, two-terminal, gas-filled tube. The voltage across it is essentially constant over a wide range of currents through the tube. It is similar in function to a Zener diode.

voltage standing wave ratio

Abbreviated *VSWR*, the ratio of the maximum voltage to the minimum voltage at a given point down a transmission line, an index of many properties of a mismatched line. Also called *standing wave ratio.*

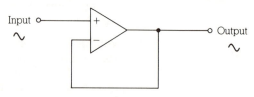

Voltage follower

voltage-to-frequency converter
See *voltage-controlled oscillator.*

volt-ampere Abbreviated *VA*, a unit of apparent power. Without taking phase into account, equal to the product of the RMS voltage and the RMS current, so that $P_{apparent} = V_{rms}I_{rms}$.

volt-amperes reactive Abbreviated *VAR*, a unit of reactive power. Equal to the product of the RMS voltage and the RMS current multiplied by the sine of the phase angle between the voltage and current waveforms, so that $P_{reactive} = V_{rms}I_{rms}\sin\theta$.

Volta's law When two dissimilar, uncharged metallic conductors are in contact with each other, the same contact potential is developed whether this contact is direct or through an intermediate conductor.

voltmeter An instrument that measures voltage, either AC or DC.

voltmeter sensitivity Expressed in ohms per volt, a measure of the loading effect of a voltmeter when placed in parallel with another resistance. It is the ratio of the total resistance of the voltmeter to its full-scale voltage reading.

volt-ohm-milliammeter Abbreviated *VOM*, a test instrument that contains the functions of a voltmeter, ohmmeter, and milliammeter in a single case.

volume density Symbolized by the Greek letter ρ, equal to the total charge per unit volume. Also called *electric charge density.*

volume unit (1) Abbreviated *VU*, a measure of the power level of an electrical signal, usually voice. For sinewaves of single frequencies, a reference of 0 *VU* equals + 4 dBm. (2) The unit of measurement defined by the zero reference level of the steady state power of 1 milliwatt dissipated by a resistance of 600 ohms at 1 000 Hz.

VOM (or vom) Abbreviation for *volt-ohm-milliammeter.*

VOX Acronym for voice-operated transmission.

VR tube Abbreviation for *voltage regulator tube.*

VSWR (or vswr) Abbreviation for *voltage standing wave ratio.*

VTVM (or vtvm) Abbreviation for *vacuum-tube voltmeter.*

VU (or vu) Abbreviation for *volume unit.*

VU meter Abbreviation for *volume unit meter.* A meter calibrated in decibels used to obtain comparative readings of speech power.

W Unit symbol for *watt.*

Wagner ground A ground connection used in conjunction with an AC bridge to minimize errors due to stray capacitance when measuring high impedances. The oscillator is removed from its usual ground connection and bridged by a series combination of a resistor and capacitor.

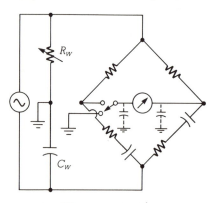

Wagner ground

watt Abbreviated W, a unit of power. The power required to do work at the rate of 1 joule per second. In an electrical circuit, the power developed when a current of 1 ampere

flows through a resistance of 1 ohm. Equal to 1/746 horsepower.

Watt, James A Scottish inventor of steam engines (1736–1819), after whom the electrical unit of power was named.

watt-hour Abbreviated Wh, the energy of 1 watt acting for 1 hour. Equal to 3 600 joules.

watt-hour meter An integrating-type meter, calibrated in terms of kilowatt-hours, that measures the accumulative active power delivered to a load over a given period of time.

wattmeter A meter that measures active electrical power in watts.

watt-second Abbreviated Ws, the energy of 1 watt acting for 1 second. Equal to 1 joule.

wave A disturbance whose magnitude is a function of time and position. A wave may be acoustic, electric, electromagnetic, mechanical, or of other forms.

waveform The graphical representation of the shape of a wave with its amplitude plotted as a function of time.

waveform generator An electric circuit that produces a wave with a specific shape, amplitude, and frequency.

waveguide A rectangular or circular pipe boundary with a specific cross-section used to guide, or conduct, high-frequency electromagnetic waves through its interior or any other equivalent system.

wavelength Symbolized by the Greek letter λ or Λ, the distance between points that have corresponding phase in two consecutive cycles of a periodic wave. The wavelength, expressed in meters, is equal to the speed of light c divided by the wave's periodic frequency f in hertz, or $λ = c/f$.

wavemeter A device for measuring the wavelength of a radio frequency wave by means of resonance effects. Since wavelength and frequency are interrelated, a wavemeter also measures the wave's frequency.

Wb Unit symbol for *weber.*

weber Abbreviated *Wb*, the unit of magnetic flux. The amount of magnetic flux, when linked to a single turn of wire, that will induce 1 volt in the turn as it decreases uniformly to zero in 1 second. Dimensionally equal to 1 volt-second. Equal to 10^8 maxwells.

Weber, William Eduard A German physicist (1804–1891) who worked with K. F. Gauss on terrestrial magnetism, introduced the absolute system of electrical units, and after whom the unit of magnetic flux was named.

Weston standard cell A highly accurate voltage source whose positive electrode is mercury, whose negative electrode is cadmium, and whose electrolyte is a saturated cadmium sulfate solution. Used as a calibration reference or standard, it has a voltage of 1.018636 volts at 20°C.

wet-cell A battery cell whose electrolyte is in liquid form and is free to move within the cell.

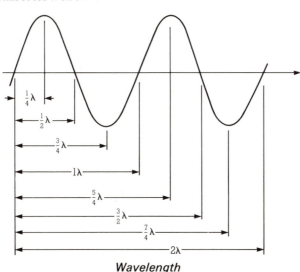

Wavelength

wet-reed relay A reed relay which has mercury at the relay contacts to reduce arcing and contact bounce.

Wh Unit symbol for *watt-hour.*

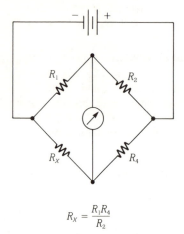

$$R_X = \frac{R_1 R_4}{R_2}$$

Wheatstone bridge

Wheatstone bridge A four-resistor arm bridge excited by either an AC or DC voltage, used for measuring resistance. Also called a *resistance bridge.*

white noise Random or impulsive noise with a flat frequency spectrum, i.e., with equal energy distribution at all frequencies.

wideband Pertaining to a network capable of passing a broad range of frequencies.

Wien bridge An AC bridge that has two adjacent resistive branches, used for the measurement of capacitance or inductance in terms of resistance and frequency. See also *Wien capacitance bridge* and *Wien inductance bridge.*

Wien-bridge oscillator An oscillator whose output frequency is controlled by a Wien bridge.

Wien capacitance bridge An AC bridge used for the measurement of capacitance in terms of resistance and frequency. It has two adjacent resistive branches and two adjacent reactive branches. One reactive branch has a resistance in parallel with a capacitor, while the other branch has a resistor in series with a capacitor. The balance depends upon frequency.

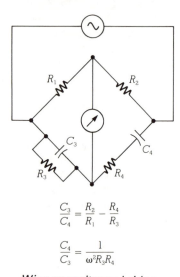

$$\frac{C_3}{C_4} = \frac{R_2}{R_1} - \frac{R_4}{R_3}$$

$$\frac{C_4}{C_3} = \frac{1}{\omega^2 R_3 R_4}$$

Wien capacitance bridge

Wien inductance bridge An AC bridge used for the measurement of inductance in terms of resistance and frequency. It has two adjacent resistive branches and two adjacent reactive branches. One reactive branch has a resistance in parallel with an inductor, while the other branch has a resistor in series with an inductor. The balance depends upon frequency.

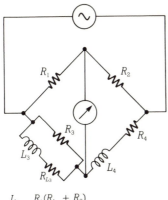

$$\frac{L_3}{L_4} = \frac{R_1(R_{L_3} + R_3)}{R_2R_3 - R_1R_4}$$

$$\omega^2 L_3 L_4 = R_4(R_{L_3} - R_3) - R_{L_3}R_3\left(\frac{R_2}{R_1}\right)$$

Wien inductance bridge

wiper (1) In a switch or stepping relay, the moving contact that makes contact with a terminal. (2) In a potentiometer, the contact that moves along the resistive element, dividing the resistance into two parts depending upon its position.

wire (1) A slender rod or filament of drawn metal. (2) A solid or stranded group of solid cylindrical conductors having a low resistance to current flow, together with any associated insulation.

wire gauge A system for classifying wire by it diameter. See *American Wire Gage.*

wire-wound resistor A resistor in which the resistance element is a length of high resistance wire or ribbon wound onto an insulated form.

wobble See *flicker.*

woofer A speaker designed to reproduce bass frequencies, or the low end of the audio spectrum.

working-Q See *loaded-Q.*

working voltage The maximum voltage rating for a capacitor.

wow A slow periodic variation in pitch, or low-frequency flutter, in sound recordings. Also called *flutter.*

wpm Abbreviation for words per minute.

Ws Unit symbol for *watt-second.*

WVDC (or wvdc) Abbreviation for working volts DC.

WW Abbreviation for wire-wound.

WWV A radio station operated by the National Bureau of Standards in Boulder, Colorado that broadcasts time and frequency standards 24 hours a day.

WWVH A radio station in Kauai, Hawaii, with services similar to WWV.

wye connection A Y-shaped winding connection with three branches. Also called a *star connection, Y-connection,* or *Y-network.*

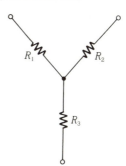

Wye-connection

wye junction A Y-shaped waveguide junction with three orifices. Also called a *Y-junction.*

X Symbol for *reactance.* Capacitive reactance is symbolized by X_C and is equal to $1/2\pi fC$. Inductive reactance is symbolized by X_L and is equal to $2\pi fL$.

X-axis The axis perpendicular to the *Y-Z* plane in a three-dimensional Cartesian coordinate system.

X-axis amplifier See *horizontal amplifier.*

X-band The radio-frequency band from 5.2 to 10.9 GHz.

xmtr Abbreviation for *transmitter.*

XOR gate Acronym for *exclusive-OR GATE.*

X-ray Electromagnetic radiation created by the bombardment of high-energy electrons onto a metal target. The resultant collision causes the atoms to lose energy, which is radiated in the form of X-rays.

X-ray spectrum Electromagnetic radiation with wavelengths from 10^{-7} to 10^{-10} cm.

xtal Abbreviation for *crystal.*

X-Y display A 2-dimensional, rectilinear (Cartesian) coordinate plot of two variables.

Y Symbol for *admittance.*

YAG Acronym for *yttrium-aluminum garnet.*

Y-axis The axis perpendicular to the *X-Z* plane in a three-dimensional Cartesian coordinate system.

Y-axis amplifier See *vertical amplifier.*

Y connection See *wye connection.*

YIG Acronym for *yttrium-iron garnet.*

Y-junction See *wye junction.*

Y-network See *wye connection.*

Y-parameters The four parameters of a two-port network, expressed as admittances so that

$$y_{11} = \frac{I_1}{V_1} \qquad \text{when } v_2 = 0$$

$$y_{12} = \frac{I_1}{V_2} \qquad \text{when } v_1 = 0$$

$$y_{21} = \frac{I_2}{V_1} \qquad \text{when } v_2 = 0$$

$$y_{22} = \frac{I_2}{V_2} \qquad \text{when } v_1 = 0$$

Also called *short-circuit admittance parameters.*

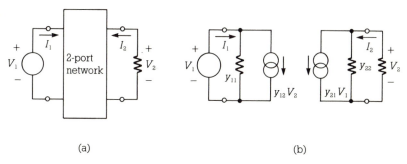

(a) (b)

Y-parameters: (a) Two-poret network; (b) Equivalent circuit

yttrium   Symbolized by *Y*, a rare-earth metallic element.

yttrium-aluminum garnet   Abbreviated *YAG*, a crystalline material sometimes used in solid-state lasers.

yttrium-iron garnet   Abbreviated *YIG*, a crystalline material used in microwave devices.

Z Symbol for impedance.

Z-axis The axis perpendicular to the *X-Y* plane in a three-dimensional Cartesian coordinate system.

Z-axis modulation The modulation of beam intensity in a cathode-ray tube. Also called *intensity modulation.*

Zener diode A two-terminal semiconductor designed to achieve avalanche breakdown at a predictable reverse bias voltage (Zener voltage).

The voltage across the diode remains essentially constant with any increase in current up to the maximum dissipation rating. When forward biased, it behaves like an ordinary rectifier. Also called a *breakdown diode.*

Zener effect The effect that is responsible for Zener breakdown in a semiconductor.

Zener impedance See *breakdown impedance.*

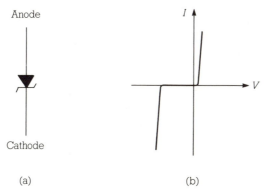

Zener diode: (a) Symbol; (b) V-I characteristic

Zener voltage The voltage across a Zener diode at which the Zener effect takes place; the breakdown voltage of a Zener diode. See *breakdown voltage.*

zero beat The condition in which a circuit oscillates at the exact frequency of another circuit, so that no beat tone is produced or heard.

zero bias (1) For a vacuum tube, the bias arrangement where the control grid and the cathode DC voltages are equal. (2) For a field effect circuit, the bias arrangement where the gate and the source DC voltages are equal.

zero-crossing detector A comparator circuit that determines if the input signal is greater than or less than a zero reference.

zero level A reference level used for comparing sound or signal levels.

zero offset voltage A zero DC output voltage from a network when the input voltage to the network is zero.

zig-zag connection In polyphase circuits, the connection of wye (star) windings, each branch of which generates a phase-displaced voltage. Also called an *interconnected star connection.*

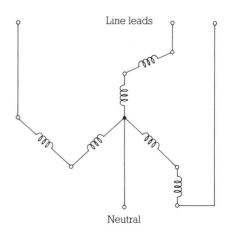

Line leads

Neutral

Zig-zag (interconnected star) connection

Z-parameters The four parameters of a two-port network, expressed as impedances, so that

$$z_{11} = \frac{V_1}{I_1} \quad \text{when } I_2 = 0$$

$$z_{12} = \frac{V_1}{I_2} \quad \text{when } I_1 = 0$$

$$z_{21} = \frac{V_2}{I_1} \quad \text{when } I_2 = 0$$

$$z_{22} = \frac{V_2}{I_2} \quad \text{when } I_1 = 0$$

Also called *open-circuit parameters.*

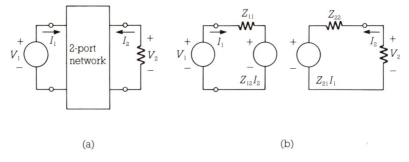

(a) (b)

Z-parameters: (a) Two-port network; (b) Equivalent circuit

RULES OF STYLE IN ELECTRICITY AND ELECTRONICS

1. Abbreviations

Abbreviations are generally written entirely using capital letters. This is done to eliminate problems with hyphenation, or when an abbreviation is the first word of a sentence.

- A hyphen is used after an abbreviation when it is combined with another word to form a compound adjective, such as in AC-coupled amplifier.
- Plural forms of abbreviations use a lowercase "s," such as *IC*s, for integrated circuits.

2. Acronyms

An acronym is a pronounceable abbreviation made from the first letter or letters of each successive word of a term or phrase. As examples, *laser* is an acronym for *l*ight *a*mplification by *s*timulated *e*mission of *r*adiation, *op-amp* is a contraction of the two-word phrase *op*erational *amp*lifier, and *LORAN* is an acronym for *l*ong *r*ange *n*avigation. As a general rule, acronyms are written in lowercase letters; however, they may either be capital letters as part of a title or in tabulation headings regardless of conventional rules for style for English writing.

3. Units of Measure

- When a letter symbol is derived from the name of a person, the first letter of the letter symbol is capitalized, such as *A* for *ampere,* and *V* for *volts* (Volta).
- When a letter symbol is made using two or more letters, only the first letter is capitalized. For example *Hz* for hertz, *Oe* for oersted. It is also permissible to spell out the unit, but be consistent in usage within a text.
- Other than when it is the first word of a sentence, title, or tabulation heading, the first letter of a unit when spelled out is not capitalized. As an example, the unit *Hz* (after Hertz), is written *hertz.*
- As letter symbols represent both the singular and plural forms of the unit, a lowercase "s" is never added to represent the plural form. As

examples, 0.23 volts is correctly abbreviated as 0.23 V, while 1 volt is written as 1 V.

• A space is used between a number and its symbol, while a hyphen is used between the number and symbol as a compound adjective; for example: a 15-kV secondary.

• Periods are used after letter symbols only at the end of a sentence and unit symbols are not used without a numerical value.

Standard prefixes used as multipliers are listed in descending order in on page 180. As a general rule, prefix magnitudes greater than 1.0 are capitalized, such as *M* for *mega* (10^6), while those less than 1.0 are written in lowercase, such as *p* for *pico* (10^{-12}).

Only two Greek letters are used to represent unit symbols, Ω for *ohm* and μ for *micro*. In lieu of using these Greek symbols, they may be spelled out but care must be used, as a vowel is sometimes omitted. As examples: microhms (micro ohms), megohms (mega ohms), etc. However, words such as microampere and kiloampere have no letters dropped. Letter symbols used in electricity and electronics are presented on page 183.

4. Numbers

• In technical writing, numbers are preferred, as they are generally easier to read than if spelled out. However, if it is the first word of a sentence, the number is spelled out.

• In a sentence, numbers under 10, and sometimes 10, are spelled out. Large numbers frequently have modifiers; for example, 1 200 000 is written as 1.2 million.

• Commas are used between every group of three figures in large numbers, starting from the rightmost figure, such as 4,784,321.05. Europeans generally use periods in place of commas and commas in place of periods, i.e., 4,784,321.05 is 4.784.321,05. A comma is optional in a four-digit number but all four-digit numbers should be treated alike. An increasing practice is to use a space instead of a comma so that 4,784,321.05 is written 4 784 321.05, which is the practice followed in this text. However, commas are not used in page numbers, year numbers, decimal fractions, or binary numbers.

• It is permissible to spell out *percent* when used with numbers, either as figures or spelled out. A space is not used between the number and the percent sign (%). As an example, six percent is written as 6%.

• With decimal numbers less than 1, a 0 is placed to the left of the decimal point because sometimes the decimal point does not show up clearly in printing. The only exception to this rule is with numbers indicating probability, which never exceed 1.0.

• Use the word *to* in place of the hyphen to avoid confusion when the upper and lower limits of a range of numbers have different units of measure, or when one or both numbers have polarity signs. As examples: − 15 to + 15 V, and 300 Hz to 3 kHz. When the upper and lower limits of a range of numbers have the same units, a hyphen may be inserted between the two numbers, but the units are not repeated. As an example, "a range from 88 MHz to 108 MHz" is written as "a range from *88–108 MHz*."

• Hyphens are used after numbers forming compound adjectives when separated by the words *and, by, or,* or *to.* As examples: 300- to 3000-Hz range, or 50- and 60-Hz line frequencies.

MULTIPLIER PREFIXES

Listed below are those prefixes used in electricity and electronics which indicate multiples and submultiples of units.

Prefix	Symbol	Multiplier	Decimal Value
exa-	E	10^{18}	1 000 000 000 000 000 000
peta-	P	10^{15}	1 000 000 000 000 000
tera-	T	10^{12}	1 000 000 000 000
giga-	G	10^{9}	1 000 000 000
mega-	M	10^{6}	1 000 000
kilo-	k	10^{3}	1 000
deci-	d	10^{-1}	.1
centi-	c	10^{-2}	.01
milli-	m	10^{-3}	.001
micro-	μ	10^{-6}	.000 001
nano-	n	10^{-9}	.000 000 001
pico-	p	10^{-12}	.000 000 000 001
femto-	f	10^{-15}	.000 000 000 000 001
atto-	a	10^{-18}	.000 000 000 000 000 001

STANDARD QUANTITY SYMBOLS

The following quantity symbols are those frequently used in the study of electricity and electronics. Where two symbols are separated by three dots, the second is to be used only where there is a specific need to avoid conflict.

Quantity	Symbol	Units
active power	P	watt
admittance	Y	siemen
angle (phase angle)	θ, ϕ	radian, degree
angular frequency	ω	radian per second
apparent power	$S \ldots P_s$	voltampere
capacitance	C	farad
capacitivity (permittivity)	ϵ	farad per meter
conductance	G	siemen
conductivity	Γ, σ	siemen per meter
coupling coefficient	κ	dimensionless
dissipation factor	D	dimensionless
efficiency	η	dimensionless
elastance	S	reciprocal farad, daraf
electric charge	Q	coulomb
electric field strength	E	volt per meter
electric flux	Ψ	coulomb
electrostatic potential	$V \ldots \phi$	volt
energy	E	joules, watt-hour
force	F	newton
frequency	$f \ldots \nu$	hertz
gain (ordinary)	A	dimensionless
gain (logarithmic)	A	bel, neper
illuminance (illumination)	E	lux
impedance	Z	ohm
inductance	L	henry
irradiance	E	watt per square meter
leakage coefficient	σ	dimensionless
luminance	L	candela per square meter, lambert, nit
luminous intensity	I	candela
luminous flux	Φ	lumen
magnetic field strength	H	ampere per meter
magnetic flux	Φ	weber

Quantity	Symbol	Units
magnetic flux density	B	tesla
magnetic flux linkage	Ψ	weber
magnetization	H, M	ampere per meter
magnetomotive force	F, F_m	ampere
period	T	second
permeability	μ	henry per meter
permeance	P, P_m, γ	henry
power	P	watt
power factor	PF	dimensionless
quality factor	Q	dimensionless
reactance	X	ohm
reactive power	$Q \dots P_q$	var
reluctance	R, R_m	reciprocal henry, ampere-turn per maxwell
reluctivity	ν	meter per henry
resistance	R	ohm
resistivity (specific resistance)	ρ	ohm-meter
susceptance	B	siemen
susceptibility	χ, κ	dimensionless
temperature (customary)	t	degree
temperature (thermodynamic)	$T \dots \theta$	kelvin
time constant	$\tau \dots T$	second
transmittance	τ	dimensionless
turns ratio	N	dimensionless
voltage (potential difference)	$V, E \dots U$	volt
wavelength	Λ, λ	meter
work	W	joule

182

STANDARD UNIT SYMBOLS

The following unit symbols are those frequently used in the study of electricity and electronics.

Unit	Symbol	Unit	Symbol
ampere	A	joule	J
ampere-hour	Ah	kelvin	K
ampere-turn	At	lambert	L
baud	Bd	lumen	lm
bel	B	lux	lx
coulomb	C	maxwell	Mx
decibel	dB	meter	m
degree (angle)	. . . °	mho	mho
degree (temperature)		neper	Np
degree Celsius	°C	newton	N
degree Fahrenheit	°F	nit	nt
kelvin	K	oersted	Oe
dyne	dyn	ohm	Ω
electronvolt	eV	radian	rad
farad	F	second	s
foot candle	fc	siemen	S
gauss	G	tesla	T
gilbert	Gb	var	var
henry	H	voltampere	VA
hertz	Hz	watt	W
horsepower	hp	watt-hour	Wh
hour	h	weber	Wb

GREEK LETTER SYMBOLS

The following Greek letter symbols are those frequently used in the study of electricity and electronics. Unless indicated otherwise, the lowercase letter is generally used.

Uppercase	Lowercase	Name	Designation
A	α	alpha	transistor forward current gain
B	β	beta	transistor short circuit current gain
Γ	γ	gamma	permeance, conductivity
E	ϵ	epsilon	capacitivity, permittivity
H	η	eta	efficiency, intrinsic standoff ratio
Θ	θ	theta	temperature, angle
K	κ	kappa	coupling coefficient, susceptibility
Λ	λ	lambda	wavelength
M	μ	mu	permeability, micro (prefix, 10^{-6})
N	ν	nu	frequency, reluctivity
Π	π	pi	constant equal to 3.14159....
P	ρ	rho	resistivity, volume density
Σ	σ	sigma	conductivity, Stefan-Boltzmann constant, leakage coefficient, summation (Σ)
T	τ	tau	time constant, transmittance
Φ	ϕ	phi	angle, potential, luminous flux (Φ), magnetic flux (Φ)
X	χ	chi	susceptibility
Ψ	ψ	psi	electric flux (Ψ), magnetic flux linkage (Ψ)
Ω	ω	omega	angular frequency, ohm (Ω)

SCHEMATIC SYMBOLS

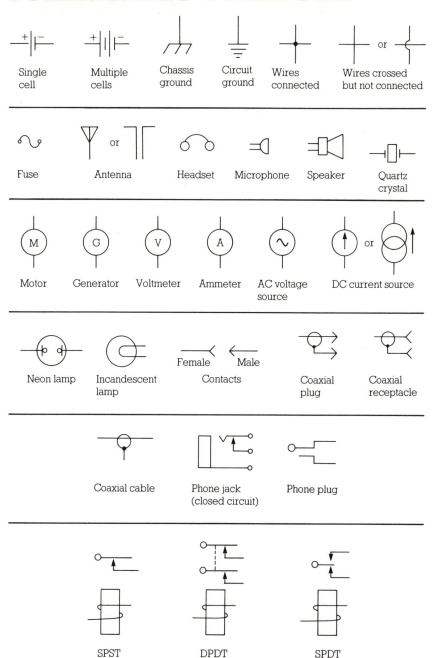

Single cell

Multiple cells

Chassis ground

Circuit ground

Wires connected

Wires crossed but not connected

Fuse

Antenna

Headset

Microphone

Speaker

Quartz crystal

Motor

Generator

Voltmeter

Ammeter

AC voltage source

DC current source

Neon lamp

Incandescent lamp

Female   Male
Contacts

Coaxial plug

Coaxial receptacle

Coaxial cable

Phone jack (closed circuit)

Phone plug

SPST

DPDT

SPDT

Relays

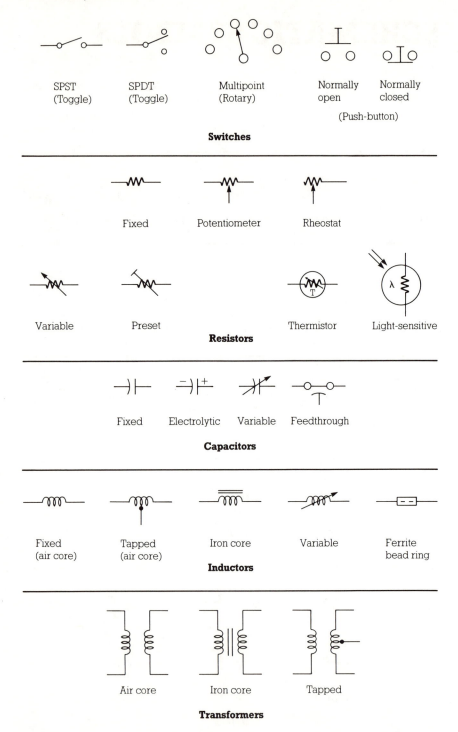

SPST
(Toggle)

SPDT
(Toggle)

Multipoint
(Rotary)

Normally
open

Normally
closed

(Push-button)

Switches

Fixed

Potentiometer

Rheostat

Variable

Preset

Thermistor

Light-sensitive

Resistors

Fixed

Electrolytic

Variable

Feedthrough

Capacitors

Fixed
(air core)

Tapped
(air core)

Iron core

Variable

Ferrite
bead ring

Inductors

Air core

Iron core

Tapped

Transformers

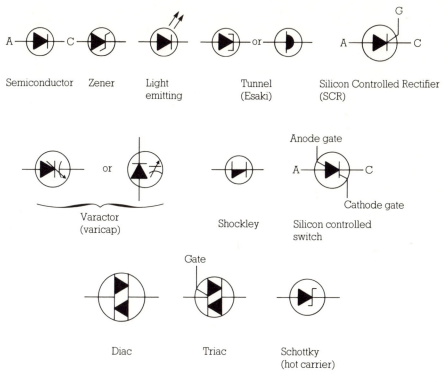

Semiconductor | Zener | Light emitting | Tunnel (Esaki) | Silicon Controlled Rectifier (SCR)

Varactor (varicap) | Shockley | Silicon controlled switch

Diac | Triac | Schottky (hot carrier)

Diodes (drawn with or without circles)

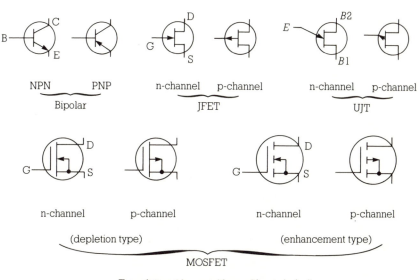

NPN | PNP | n-channel | p-channel | n-channel | p-channel
Bipolar | JFET | UJT

n-channel | p-channel | n-channel | p-channel
(depletion type) | (enhancement type)

MOSFET

Transistors (drawn with or without circles)

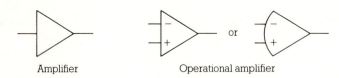

Amplifier   Operational amplifier

Amplifiers

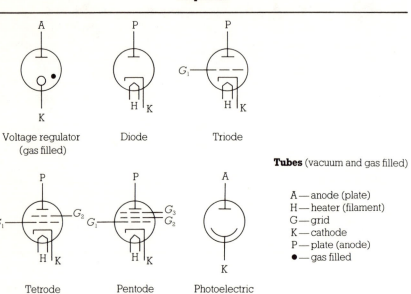

Voltage regulator   Diode   Triode
(gas filled)

Tetrode   Pentode   Photoelectric
cell

Tubes (vacuum and gas filled)

A — anode (plate)
H — heater (filament)
G — grid
K — cathode
P — plate (anode)
● — gas filled